DATA
WHERE IT IS AND HOW TO GET IT

The 1993 Directory of

BUSINESS ENVIRONMENT and ENERGY

Data Sources

Edwin J. Coleman and Ronald A. Morse

DATA: Where It Is and How To Get It

by Edwin J. Coleman and Ronald A. Morse

Illustrations by Ernest Kellett

All Rights Reserved

Copyright © 1992 by Coleman / Morse Associates Ltd., all rights reserved. No part of this book may be used or reproduced in any manner whatsoever without written permission from the publisher except in the case of brief quotations embodied in critical articles, reviews, bibliographies, and scholarly works.

The publisher makes every effort to accurately report information supplied by public sources, but is not responsible for and disclaims any liability for information inaccurate at its source or for changes that result from the passage of time.

Printed in the United States of America

ISBN 0-941375-56-0

For information contact:

Coleman / Morse Associates Ltd.
Suite 231
1290 Bay Dale Drive
Arnold, MD 21012-2325

Preface

We see this book as our modest contribution to America's global competitiveness and business productivity. We feel strongly that the time has come for Americans to become more data literate. That is the primary reason we have prepared this hands-on, practical guide to useful business, environmental and energy data.

This book meets an important need. For years, we have seen professionals in all walks of life make important business and political decisions without using the best data available. We know that their effectiveness was limited as a consequence. To be sure, there are explanations for their behavior. Often they didn't know whom to ask for good data and that is understandable. More frequently, these people were afraid to admit that they were uncomfortable with data and therefore opted to ignore it. This too is unfortunate, but it is understandable. The basic problem was that they didn't know where to turn for help. Up to now there has been no accurate and easy-to-use guide for non-economists about how to get, understand and use business-related data.

We feel we have remedied these problems in this one volume. Half of the book is instructional. It is an introductory guide to data—where it is produced, how it is prepared, the tricks to using it and the jargon that seems to make it difficult to use. We call this section the **DATAPRIMER** because it explains everything anyone needs to know to feel comfortable with data and to use it effectively. We have made every effort to be clear and practical, without sacrificing accuracy.

The other half of the book contains three practical and well indexed directories to business, environmental and energy data. This is the **DATAPHONER** section of the book and it is the heart of the volume. It contains the names and areas of specialization of over 2,500 individuals in federal, state and local governments, and in private firms who can answer specific questions about nearly every business-related data issue of interest to busy professionals.

We also wanted the book to be a practical working aid. We could have produced another large and expensive directory that would have just sat on the office bookshelf, but instead we selected those data experts who are producing the economic, financial, demographic, environmental and energy numbers. We have organized the data into a series of easy to use directories. This way we could also make the book handy and affordable. This **1993 Directory of Data Sources** is yours. Take it with you on trips, scribble notes in it and update it as necessary. It puts you directly in touch with the data collectors and compilers who have the information you need to do the job right. You also have the satisfaction of knowing that it is the most up-to-date directory of its kind.

This book should be useful to a broad spectrum of people. It is most obviously of immediate practical use to economists and corporate groups that use business, environmental and energy data on a daily basis. Also, scholars and think tanks will find it a practical tool. It can also be used as a text in business schools and economics courses to train students in the use of data. Consultants and others will also find it exceptionally valuable in their work.

We think we have done our best in preparing the book, but we may have missed something. Or perhaps we included something that turns out not to be very useful. Please let us know. We welcome your ideas, comments and suggestions for the 1995 edition.

Edwin J. Coleman
Ronald A. Morse

Contents

Preface	i
Introduction	vii
Part One. The DATAPRIMER	1
I. Getting Started: The Art of Using Data	3
II. The Data Makers	7
The "Big Four" Economic Data Factories	7
The Major Producer of Banking and Financial Data	8
Other Important Data Factories	9
III. The National Trade Data Bank	13
IV. Commercial Data Services	15
V. Good Data, Bad Data: How to Tell the Difference	19
VI. Making Estimates	23
VII. The Best Business-Related Data Publications	25
Economic and Financial Data Publications	27
National and International Statistical Guides	29
Educational Data Publications	31
The Best Summary of U.S. Statistics	33
Government Printing Office Bookstores	34
VIII. Jargon: Terms, Concepts and Measures	37
Commonly Used Economic and Statistical Terms	39

Contents

    IX. Using Statistical Tables 55

    X. Additional Readings on Data Issues 59

Part Two. The DATAPHONER 61

    XI. Business Data Directory 67

        Section I. Business Data Sources 69
                Agricultural Data 71
                Banking and Financial Data 78
                Business and Industry Data 83
                Data Managers 88
                Data Products and Other Information Services 91
                Employment and Labor Force Data 94
                Federal, State and Local Government Data 96
                Geographic Concepts and Products 97
                Housing Data 98
                International and Foreign Trade Data 99
                Personal Income and Related Data 103
                Population and Other Demographic Data 105
                Price and Price Index Data 107
                Rural Development Data 109
                Statistical Concepts and Methods 110
                Unemployment Data 112
                Working Conditions and Compensation Data 113

        Section II. State and Regional Data Centers 115
                The Census Bureau: State Data Centers 117
                The Census Bureau: Business and Industry Data
                    Centers 123
                The Bureau of Economic Analysis: State User Groups ... 127
                The Bureau of Labor Statistics: Regional Offices 153
                The National Agricultural Statistics Service State
                    Statisticians' Offices 159

        Section III. Commercial Data Services 165

Contents

XII. Environmental Data Directory	183
Air Quality	191
Land Use	193
Solid and Toxic Waste	196
Water Quality and Water Use	197
Other Policy Relevant Environmental Data	201
XIII. Energy Data Directory	203
Section I. Energy Data Sources	207
Coal and Coke	209
Crude Oil	209
Electric Plants and Electric Utilities	210
Natural Gas	210
Oil and Petroleum Products	211
Other Energy Sources	211
Section II. Uses of Energy Data	213
XIV. Index of Data Experts	223
Subject Index	249
About the Authors	257

Introduction

Everyone knows that information is power. What they don't know, and that is why this book is important, is that information in the form of hard primary data is extremely powerful. This book is your guide to that data—the critical data knowledge you need to come out on top in a competitive world. The **1993 Directory of Data Sources** is your data power weapon—it is your personal guide to hundreds of millions of dollars worth of official business, financial, environmental and energy data. Telephone information lists, compiled by subject in a format not available anywhere else, take you right to the people who control the vital information on nearly every aspect of activity.

But you ask, is this book just another business rolodex? The answer is no! It is a data guide that avoids the pitfalls of the usual unmanageable—everything on the subject—type of guide. The **1993 Directory of Data Sources** is a primer on everything you need to know to identify, find and use business-relevant data. The core of this capability is three data directories—a business data directory, an environmental data directory and an energy data directory. Each directory is organized to give you the data items that have the most in the way of practical value. We have taken the time, so that you don't have to, to evaluate the available sources and we have selected only those items that we think will be most useful to you.

If you are a little uncertain about using data, you will want to begin with Part One of the book, the **DATAPRIMER**—a professional and user-friendly guide to using and understanding data. Reading this before turning to the data directories is well worth the time. We have found that going through the **DATAPRIMER** is even useful to people who have quite a bit of experience using data.

It is no exaggeration to say this book is one of a kind. It opens doors to information that budget cutters in Washington have tried to deny to the very people who have paid for it—you the taxpayer. For years, insider data pros have had direct access to the data items and experts listed in this volume. Now this direct access is available to you.

Introduction

Closing the Data Gap

Smart people know that the secret to making good business decisions is closing the **data gap**—finding the data they need and then spotting any gaps between what the data say is happening and what is really happening. Everyone needs data that are reliable and that accurately reflect what is happening. Because we know where these data are, how to get it, and how to compile it into a focused and useful format, the critical data you need for your work are here at your finger tips—everything you need to locate over 2,500 data items and data specialists is in this one handy volume.

The **1993 Directory of Data Sources** will pay for itself over and over again. It was prepared by data users for data seekers. Ed Coleman is one of the nation's leading experts on regional economic data and he knows what business and professional people need in the way of hard data. He headed up a major statistical program for 20 years and he knows the inside story about government economic data and how to use it. Ron Morse is one of the nation's leading experts on Japan and he knows that one of the secrets to Japan's success has been effective data collection and distribution. The Japanese government does just the opposite of the U.S. government—it makes sure that business has the best data money can buy. This book tries to do for you what your government won't do—it gives you access to the data necessary to be successful and competitive.

A quick glance through the pages of the book will give you an idea of just how comprehensive it is. This is a "Big Little Book"—big in terms of coverage, little in terms of size. It is full of useful business-related information, but compact enough to fit into your briefcase. It is also easy to use because it is well indexed. Most important, it puts you back in control. You don't have to worry about calling your staff or hiring expensive consultants to track down the numbers you need. If there is a telephone near by, with the information in this book, you have a direct line to the best American data sources available. We have also included tips for talking to the data pros, a handy glossary of the most useful technical jargon—economic terms, measures and statistical concepts.

Over 2,500 Data Contacts

The **DATAPHONER**, has three directories and links you with over 2,500 data sources and experts across the nation. Each directory contains the most authoritative contacts for current and practical business-related data.

The first part, the **Business Data Directory** represents the core of the U.S. government's economic data network. It consists of three sections: Business Data Sources; State and Regional Data Centers; and Commercial Data Services. The State and Regional Data Centers section and the

Introduction

Commercial Data Services section are your guides to state and regional data—the names, telephone numbers and locations of more than 800 business, government and academic contacts in every one of the 50 states and the District of Columbia. These state data centers and commercial data service suppliers can provide state and local data, customize business-related data to meet your specific requirements, and answer questions on local data issues.

Next, the **Environmental Data Directory** lists the contacts for the most frequently sought after business-relevant environmental statistics.

The **Energy Data Directory** will put you in touch with the data managers of the business and environmentally significant energy data. It consists of two sections: Energy Data Sources and Uses of Energy Data.

This book contains nearly everything you will need in the way of practical information about data. You get key customer service contacts for the government's most important Data Factories, information about how to order press releases, publications, computer tapes, diskettes and other information services, including how to access data bulletin boards. And to facilitate your access to economic publications, the **1993 Directory of Data Sources** includes the telephone numbers and addresses of the Government Printing Office bookstores around the nation. It also guides you to the best data publications and includes a bibliography for further reading.

Now, let's get busy learning about data.

PART ONE
THE DATAPRIMER

Getting Started: The Art of Using Data

The Data Makers

The National Trade Data Bank

Commercial Data Services

Good Data, Bad Data

Making Estimates

The Best Business-Related Data Publications

Jargon: Terms, Concepts and Measures

Using Statistical Tables

Additional Readings on Data Issues

I. Getting Started: The Art of Using Data

Good data are hard to find these days because over the last ten years the White House and the Congress have either cut or limited funding for the collection, analysis and dissemination of data. This has led to bureaucratic data wars pitting dedicated public servants, who know the importance of good data, against penny-pinching politicians with no long-term commitment to providing the data needed by industry, government and the public. One casualty of these wars has been the loss of data and more importantly the loss of direct access to government data experts.

It hasn't always been this way. Up until the late 1970s, the U.S. statistical data system was the best in the world. But today, because of neglect, cuts in funding and the deterioration in reporting requirements, to say nothing of the growing complexity of calculations due to global economic interdependence, current economic data are not meeting the growing needs of the American business community. Even foreign business groups complain that they have to invest a great deal of time and money to get accurate data about the U.S. economy.

This data crisis has been ignored, but it has not gone unnoticed. Over twenty official studies have identified the improvements that are needed in the way America collects its data. The issue has been brought to the attention of the White House and the Congress on numerous occasions. Since 1984, more than a hundred professional articles have been published dealing with the failures of U.S. statistical policy. Only recently, in the face of growing economic problems, have politicians recognized the importance of funding business-related statistical programs. All of this complicates the issue of data accuracy and makes it essential that users be able to locate data that they can use with confidence.

That is why the **1993 Directory of Data Sources** is so important—data users do not have the easy and reliable access to business-related data they once had. Currently, the government simply isn't providing data users with enough current information about the nation's economy. Also, it is difficult to know whom to contact in Washington for detailed information about data. For all of these reasons, it is more important than ever for you to consult directly with the nation's data experts. The **1993 Directory of Data Sources** restores your direct access to many of the nation's most knowledgeable data experts.

Part One: The DATAPRIMER

Use the Data! You Paid For It.

Even though Washington does not have a coherent policy on data, your tax dollars are being used wisely by the data pros—the dedicated federal economists and statisticians still busy collecting the data you need. In the pages that follow, we describe the output of the government's Data Factories, we discuss the quality of data sources and estimates and give you the capability to tap directly into that valuable pool of knowledge using the data directories and indexes.

Knowing how the government's record-keeping system for business-related data works can help you get an edge on your competition at home and overseas. Since U.S. statistical methods and concepts are widely imitated by other nations, the more you know about your own American data system, the more you will know about your foreign competitors as well.

Fortunately, you can get most of the data you need from what we call the Data Factories—the key government statistical agencies described here. By using the directories in the **DATAPHONER**, you will learn who in which Data Factory has the data you need. And we have set up the directories so that it will be easy for you to get these data over the telephone. Forget the costly travel expenses and outside experts. Just let your fingers dial your way to success.

Data Factories Ready to Serve You

As with any product, if you want to know how good the quality is, you have to go to the shop floor and check out how it is made. With data that means knowing something about how and where the data are assembled, adjusted, tabulated and turned into a finished product. What the Data Factories produce is explained in detail in the next section. Here, we will just give you some background information about them.

To start, see if you can guess where the Data Factories are located? That's right. The largest concentration of Data Factories is in Washington, DC—that's right, in and around the heart of the nation's Capital. And when it comes to official numbers, good old Uncle Sam, despite all his failings, is still the only source for much of the data you want.

Until 1982, the government had a separate office to coordinate all of its federal statistical programs. That unfortunately is no longer the case. Now 70 scattered Data Factories are coordinated (badly, one might add) by the Statistical Policy Office in the Office of Information and Regulatory Affairs of the Office of Management and Budget.

I. Getting Started: The Art of Using Data

In 1935, the Federal government started publishing the **Federal Statistical Directory** as one in a series of publications for the "orderly dissemination of information about Federal statistical activities." In the late 1970s, budget cutters in Washington discontinued the practice of giving the people who pay for the data—you, the taxpayer—convenient access to the data. The Federal Statistical Directory is not published any more. The **1993 Directory of Data Sources** restores your direct access to a broad range of important business-related data experts.

That's the big picture. Now, let's look more carefully into the Data Factories and the data they produce.

Federal Depository Libraries. There are 1,400 libraries scattered around the nation that receive a variety of federal publications. Many of these publications are reports and publications from key agencies that prepare business-related data.

Data Customer Service Reps are a Good Place to Start!

If you are not exactly sure what data you need or which government agency produces it, contact one of the customer service offices listed below. If they don't have the data you need, they will point you in the right direction.

If you know what data you need, but have questions about the reliability, availability or methodology used to produce the data, contact the data experts listed in the **DATAPHONER** directories.

** **U.S. Department of Agriculture,** Economic Research Service and National Agricultural Statistics Service, Information Division, **(202) 219-0494**

** **U.S. Department of Commerce,** Bureau of the Census, Customer Services, **(301) 763-4100**

** U.S. Department of Commerce, Bureau of Economic Analysis, Office of Public Information, **(202) 523-0777**

** **U.S. Department of Energy,** Energy Information Administration, National Energy Information Center, **(202) 586-8800**

** **U.S. Department of Housing and Urban Development,** Housing and Demographic Analysis Division, **(202) 755-5630**

** **U.S. Department of Labor,** Bureau of Labor Statistics, Publications Office, **(202) 523-1206**

** **U.S. Department of the Treasury,** Internal Revenue Service, Statistics of Income Division, **(202) 376-0216**

** **U.S. Environmental Protection Agency,** Center for Environmental Statistics, **(202) 260-2680**

** **Federal Reserve System,** Board of Governors, Publications Office, **(202) 452-3244**

II. The Data Makers

In this section and the two following it, we introduce the important producers of data. The next section introduces the National Trade Data Bank. Section IV familiarizes you with the array of services offered by commercial data firms.

To begin with there are five major economic and financial Data Factories. Four are officially called statistical agencies, the fifth is the Nation's central bank. Together they produce nearly 90 percent of the national and regional business-related data that you will ever need. Each year the federal government's statistical agencies, offices and departments spend more than $1,000,000,000— that's right, a billion dollars of the taxpayer's money—to put together one of the most comprehensive systems of economic, financial and demographic data available anywhere in the world.

The "Big Four" Economic Data Factories

** The Bureau of the Census (Census) conducts the periodic five-year economic censuses, the censuses of government and agriculture, and produces data on construction, manufacturing, retail and wholesale trade, services, foreign trade, state and local government finances and employment.

** The Bureau of Economic Analysis (BEA) draws on the information of more than 400 reports, surveys and tabulations produced by 42 other agencies. It produces the national income and product estimates. It also produces data on input-output accounts, balance of payments and foreign investment accounts, as well as personal income and related economic estimates.

** **The Bureau of Labor Statistics (BLS)** produces data on employment and unemployment, prices and living conditions, consumer expenditures, wages and employee benefits, industrial relations, productivity and technological changes as well as projections of economic growth, labor force estimates and employment by industry and occupation.

** **The National Agricultural Statistical Service (NASS)** produces data covering domestic agriculture, including estimates of production, stocks, inventories, prices, disposition, utilization, farms and farm land as well as other related data. It is the major data source for the U.S. Department of Agriculture's **Economic Research Service (ERS)**.

> **WHAT IS ERS?** ERS is the primary analytical organization of the USDA. Its principal functions include research, situation and outlook analyses, and the development of economic and statistical indicators. ERS produces periodic reports that analyze the current situation and forecasts the short-term outlook for major agricultural commodities, agricultural exports, agricultural finance, agricultural resources and world agriculture.

The Major Producer of Banking and Financial Data

** **The Federal Reserve System**, which is managed by a Board of Governors, is the central bank of the United States. It produces information on just about every aspect of banking and finance, including data ranging from agricultural credits and automobile loans to international capital flows and thrift acquisitions.

It is the responsibility of the Federal Reserve System to contribute to the strength and vitality of the U.S. economy. By influencing the lending and investing activities of depository institutions and the cost and availability of money and credit, the Federal Reserve System can promote the full use of the Nation's human and capital resources, the growth of productivity, relatively stable prices and equilibrium in the Nation's international balance of payments.

The Federal Reserve Board in Washington and each regional federal reserve bank has its own individual statistics and research department staffed with data experts who systematically gather banking, financial and other business-related data and issue analytical reports on various aspects of the economy. These data and reports are published in the Federal Reserve Bulletin and in the regional federal reserve banks' publications.

II. The Data Makers

Other Important Data Factories

In addition to the 90 percent of the business-related data provided by the Big Four Data Factories and the Federal Reserve System, the other 10 percent of the data can be found in Data Factories affiliated with the statistical offices of larger agencies. The bulk of the data produced by these Data Factories, like that produced by the Federal Reserve System, are the byproduct of administrative or regulatory activities. The principal customers for the output of these Data Factories are the parent organizations or other agencies. Several of the more important business-related Data Factories are:

** **The Energy Information Administration (EIA)** of the U.S. Department of Energy produces data on energy resources, production, distribution, and consumption. EIA prepares reports on fuel sources and maintains statistical systems on international energy supply, demand balances and economic and financial data.

** **The Housing and Demographic Analysis Division** of the U.S. Department of Housing and Urban Development produces the annual Housing Survey. It provides data series on national, regional and local economic and housing market conditions. It has information on the physical and financial characteristics of national and selected metropolitan housing inventories and on the characteristics of occupants, housing units under construction or completed, new one-family home sales, market absorption of new rental apartments and condominiums, the placement of new mobile homes and mortgage lending and commitment activity.

** **The Statistics of Income Division** of the Internal Revenue Service (IRS) provides annual income, financial and tax data based on individual and corporate income tax returns. The Statistics of Income Division also releases periodic studies based on returns filed by estates and trusts and does in-depth analyses of tax-related computations, including foreign tax credits and sales of capital assets.

** **The Office of Business Analysis** of the U.S. Department of Commerce is a central source for trade data focused on international trade as well as data useful to U.S. business firms engaged in export-related activities. The trade data are from the National Trade Data Bank described in the next section.

*The annual **Census Catalog and Guide** not only gives you detailed information about data products and services, but also includes a detailed breakdown of the data products and publications of other government Data Factories. For more information write: Customer Services, Bureau of the Census, Washington, DC 20233, or call (301) 763-4100.*

Part One: The DATAPRIMER

The Data Factories Are Useful to Everyone

Big and Small Businesses

Businesses compare their own company statistics with economic data for their industry or area to:
- Compute market share.
- Evaluate their own growth.
- Evaluate product lines relative to the competition.

Firms that make or sell products used by other businesses use economic data to identify their markets and apply this information to:
- Map out sales territories.
- Allocate funds for advertising.
- Locate plants, warehouses or stores.
- Make sales forecasts.

Manufacturers and distributors use economic data to pinpoint the location of retailers, wholesalers, contractors, and others who may redistribute their products.

Small businesses use economic data describing their markets or their industry to enhance presentations to bank officers or venture capitalists when they seek financing.

Entrepreneurs and Independents

Easy access to information is the key to creative business ventures.

Trade Associations and Professionals

Trade associations alert their members to statistical trends affecting their industry, such as changes in domestic production relative to imports.

Professional associations use economic data to adjust the results of their own surveys to be more representative.

Law firms need data to represent clients.

Think tanks rely on business-related data in their research.

The Media

The media rely on selected statistics as background for articles and reports.

Newspapers, TV and radio use economic data in their reports.

Universities and Students

Universities use data to support their research and keep their faculty informed. Students facing career choices need a broad range of data.

Source: U.S. Department of Commerce, Bureau of the Census

II. The Data Makers

Contacts for Major Data Factory Publications

For **Federal Reserve System** publications write to: Publications Services, Mail Stop 138, Board of Governors of the Federal Reserve System, Washington, DC 20551, or call **(202) 452-3244** or **(202) 452-3245**.

For **Bureau of the Census** publications write to: Customer Services, Data User Division, Bureau of the Census, U.S. Department of Commerce, Washington, DC 20233, or call **(301) 763-4100**.

For **Bureau of Economic Analysis** publications write to: Public Information Office, BE-53, Bureau of Economic Analysis, U.S. Department of Commerce, Washington, DC 20230, or call **(202) 523-0777**.

For **Bureau of Labor Statistics** publications write to: Inquiries and Correspondence Branch, Office of Publications, Bureau of Labor Statistics, Washington, DC 20212, or call **(202) 523-1221**.

For **National Agricultural Statistical Service** and **Economic Research Service** publications write to: Information Division for NASS and ERS, Room 228, 1301 New York Avenue, N.W., Washington, DC 20005, or call, toll free, **(800) 999-6779**.

To obtain other official U.S. government data publications write to: Superintendent of Documents, U.S. Government Printing Office, 710 North Capitol Street, N.W., Washington, DC 20401, or call **(202) 275-2091**.

*The **National Technical Information Service (NTIS)**, an agency of the U.S. Department of Commerce, is the distribution center for business-related data as well as for most other technical information produced by the U.S. Government. This information and sales-oriented government agency manages the Federal Computer Products Center, which provides the public and the government with access to software, data files, and databases produced by all of the government's Data Factories. For a free NTIS Catalog call: **(703) 487-4650**, and ask for the "**NTIS Products & Services Catalog, PR-827**."*

III. The National Trade Data Bank

If you are interested in trade data, you need to know about the **National Trade Data Bank (NTDB)**. The NTDB is an ambitious and relatively new source for trade information created by the Congressional 1988 Omnibus Trade and Competitiveness Act. NTDB provides electronic access via read-only-memory compact disks (CD-ROMs), an electronic bulletin board and magnetic tapes. The primary medium for distribution of NTDB data is the CD-ROM. NTDB CD-ROMs containing the entire data bank are updated monthly.

The National Trade Data Bank is located in the **Office of Business Analysis of the U.S. Department of Commerce.** For information on how to get trade data call **(202) 377-1986.**

The NTDB Has Two Key Components

The NTDB's International Economic Data System includes data on imports and exports; international service transactions; international capital markets; foreign direct investment in the U.S.; labor markets and foreign government policies affecting trade; import and export data on a state-by-state basis aggregated at the product level; and data for countries with important economic relations with America.

The NTDB's Export Promotion Data System includes data on the industrial sectors and foreign markets of greatest interest to U.S. business firms engaged in export-related activities—specific business opportunities in foreign countries; specific industrial sectors in foreign countries with high export potential; significant tariff and trade barriers and other laws and regulations regarding imports, licensing, and the protection of intellectual property; export financing information, including the availability through public sources of funds for U.S. exporters and foreign competitors; and transactions involving barter and countertrade.

IV. Commercial Data Services

The U.S. government agencies we have been describing provide a limited number of data services: releases of unpublished data, special surveys and tabulations, and the preparation of analytical software. Commercial data services, on the other hand, often package government data in a format that can meet specific analytic or market requirements. The quality of data provided by private firms varies greatly. Some firms maintain standards as high as or higher than those of the top government Data Factories. Other firms are less reliable, honoring any data gathering request with little or no regard to the reliability or usefulness of the final product.

Be sure you are precise about what you want from commercial data firms—they package data in their own way, mixing a variety of public and private data sources. If a firm promises that their commercial data are more current or include more components or more geographic detail than that produced by a government agency, check it out thoroughly. Find out how the data were put together and ask where they got their data. Remember, when it comes to estimates of economic activity, you only get what you pay for. There are no short cuts to good data. Also, remember, it takes a lot of time and money to collect and compile data that accurately reflects economic reality.

Bad data can result in costly miscalculations. One way to check on what you are really getting from a commercial data-merchant is to ask to see the methodology they used to gather, manipulate and produce the data you requested. Examine it as carefully as you would government documentation and don't allow the commercial supplier to be vague.

A good way to find reliable commercial data sources is to check with the commercial data services identified by the Census Bureau's National Clearing House—see the third section of chapter XI, the Business Data Directory in the DATAPHONER part of this book. This is a useful source for services offered by business, government and academic organizations. The Census Bureau suggests that if you need data in machine readable form, or if you need customized tabulations, you may find it useful to consult the businesses listed by the National Clearing House.

Part One: The DATAPRIMER

When in doubt, do what the professionals do—assess the quality of data produced by a commercial Data Factory by checking out the methodology used to produce the estimates. We give some hints on how to do this later in the book. Many commercial data sources, like their government counterparts, publish or make available descriptions of their sources and methods of estimation. If these statements are not available or are considered proprietary (therefore confidential) you should again do what the professionals do—approach the commercial data with skepticism or look elsewhere for more reliable data.

*Don't overlook **Trade Associations** as a source for industrial data. Trade associations are major producers of industrial data. Keep in mind that the reliability of trade association data depends upon the ability of the trade association to get good data from its members. As a general rule, approach trade association data with the same caution that you would any other private (or government) data source.*

IV. Commercial Data Services

Data Services Provided by Commercial Firms Participating in the Census Bureau's National Clearinghouse Program

General Services

Tape copies, printouts, extracts
Software
Personal computing services
Diskettes, downloading, etc.
Online data services
CD-ROM products and services
Estimates and projections
Tabulations from microdata
Market segmentation
Market research services
Training and consultation

Data Specialization

Demographic data
Socio-economic data
Economic—retail and foreign trade

Data Holdings

National data
Regional data
Local data

Geographic Services

Geocoding and address matching
Mapping and cartography
Business graphics
Redistricting services
Routing or delivery assistance
Training and consultation
Geographic information systems

Other

Software development
Newsletters / technical journals

NOTE—The Census Bureau, and to a limited degree the other major data factories, can perform specific services including designing and carrying out sample surveys, providing population and income estimates and projections, and offering technical assistance.

The Census Bureau can be especially useful if you need to do computer mapping, generate routing plans or create a geographic information system through its **TIGER/LINE** files. The **TIGER/LINE** system is an electronic mapping program consisting of geographic computer files with latitudinal and longitudinal coordinates for the line segments of every block of land in the United States.

V. Good Data, Bad Data: How to Tell the Difference

Not all data are reliable and being able to tell the difference between good data and bad data can be useful. "Data" according to Webster's New Collegiate Dictionary is "factual information used as a basis for reasoning, discussion or calculation." But this doesn't tell us very much. If we stretch Webster's definition a little, data would include: information about something known or about something assumed; information as text, numbers or graphics; and information in any form that has meaning.

And even though many people assume data to be factual, that is not always the case. Data can be a best estimate of reality, an assumption about reality based upon incomplete information or (sometimes) poor judgment. In some instances, what appears to be factual is no more than an educated guess.

There are three basic questions economists and statisticians ask themselves when evaluating the quality of economic or other business-related data: (1) What's the size and frequency of the revisions to the data? (2) How valid is the logic of the methodology used to produce the data? and (3) How was the source data generated and how closely does it match the economic activity in question? Let's examine each of these questions.

The Size and Frequency of Revisions

The size and frequency of revisions is a common, but complex issue. For example, if you rely on the size and frequency of revisions to assess the quality of a data series without looking at other factors, you can't assume that the smaller the monthly, quarterly or annual revisions, the better the data. Small revisions or no revisions in the data do not necessarily mean that the data are reliable. Small or no revisions may mean (a) more current or comprehensive source data are not available, (b) more current or comprehensive source data are only available at periodic intervals, (c) there is no way to improve on the current methodology, or (d) the resources needed to introduce improved data sources and methods are not available due to fiscal or operational constraints. But, if your data contact cites reliable data sources and is forthcoming about the methods of estimation, and the methods appear logical or reasonable, it is safe for you to assume that the data are reliable. If revisions over time are frequent and large, the data series, of course, should be considered unreliable.

Methodology: A Major Key to Data Quality

Most professionals consider published descriptions of sources and methods used by a Data Factory to be the only foolproof means of assessing the quality of the numbers. This is because published descriptions (1) contain details about the source data used, (2) spell out the procedures used to fill in data gaps, (3) define statistical and definitional adjustments, and (4) generally include a detailed explanation of modifications based on value judgments by the economist or statistician.

When asked, most government Data Factories will provide a written explanation of how their estimates for a specific economic activity were made. If a government Data Factory does not have a written explanation of its sources and methods available for distribution, the careful data user should talk to someone with hands-on knowledge about the information needed.

If you run into this problem flip to the **DATAPHONER**, look up the subject you are interested in, pick up the telephone and call the data contact. Ask your contact to tell you what sources and methods were used to produce a particular economic measure.

The Sources of Data

This issue frequently involves assessing the quality of data produced by Data Factories that generate their estimates from other data sources. For example, the Bureau of Economic Analysis, which is responsible for the national income and product accounts, and the Federal Reserve Board, with its index of industrial production, are Data Factories that operate in this manner. Nine times out of ten, the source data used will be generated by a mix of administrative records, regulatory reports or business establishment surveys.

All three of these data sources—administrative, regulatory and business establishment—are subject to processing errors as data are tabulated and compiled. Administrative record and regulatory data will range from good to excellent and both are available on a regular basis. Administrative and regulatory data are also more likely to be comparable over time than business establishment survey data. All three types of source data can be, and often are, limited by confidentiality restrictions. Because these three types of source data often do not coincide, either statistically or definitionally with the activity being measured, they frequently require substantial modification.

Administrative Record Data

Administrative record data are the by-product of the administration of federal, state and local tax programs, social security programs, unemployment insurance programs and other federal, state and local programs. The data quality ranges from good to excellent. The data are generally available on a more frequent basis than data from surveys or other data sources. Because administrative records generate measures of activity as by-products of some other function, they may require substantial modification before they can serve as a reliable measure of an activity.

Regulatory Report Data

Regulatory data are based on government regulatory reporting requirements. With the deregulation of energy, transportation, banking and other industries over the past decade, the amount of this information generated as a by-product of government regulations has declined sharply. The downside to data generated from regulatory reports is that it is frequently unpublished and difficult and expensive to compile.

Establishment Survey Data

Establishment (industry and business) survey data depend on the ability of businesses to provide the Data Factories with the information needed to measure the changes in specific sectors of the U.S. economy. Establishment surveys are used to produce data that are not easily obtained from other sources. Unlike other more widely used types of surveys, there are few commonly accepted approaches to the design, collection, estimation, analysis, and publication of the results of establishment surveys. These data are frequently used to replace economic data lost as a result of budgetary or policy (deregulation) decisions. They include business information on employment and wages, sales, prices, agricultural production, the money supply, and many other aspects of the economy. The quality of the data produced by business establishment surveys depends less on what is included, and more on what is overlooked or not included—what the data pros call "nonsampling errors"—specification, coverage, response, and processing errors.

Why Data Need To Be Examined Carefully

Less government information is now available because of reduced funding and federal deregulation. Deregulation, which may reduce or eliminate regulatory reporting requirements, has hit communications and transportation data the hardest. The quality of some surveys has been compromised because reporting requirements are no longer mandatory.

Funding is often not available to develop reliable alternative sources to replace discontinued data series. As a result it is frequently necessary to substitute less reliable data based on related or secondary sources for lost data. Some series may be linked to newer measures which are not fully comparable.

There are increasingly frequent delays in the release of economic data due to slowdowns in the collection and processing of source data.

Accuracy has been compromised due to a lack of staff to edit and refine the data. Because of this reduced capability, poorly edited records reduce the reliability of data.

Comprehensiveness has been reduced because data formerly included in surveys have been discontinued or truncated.

The demand for current information is forcing statistical agencies to divert resources from producing data on current business activities to turning out forecasts based on fragmentary data.

VI. Making Estimates

Estimates are statistical pictures of an activity based on data that can be, and often are, a mix of administrative records, regulatory reports and business establishment survey data. In preparing an estimate, it is often necessary to make adjustments to the source data because of differences in data definitions, gaps in the data or deficiencies in data coverage. The amount of source data processed can be very large, requiring the estimator to use a variety of electronic and related editing procedures to catch errors in the source data as well as errors resulting from the estimating procedures.

Estimates can be the product of detailed and complex sets of assumptions and calculations. A good example of a complex and sophisticated economic estimate is the nation's Gross Domestic Product, the total value of the goods and services produced in the economy. On the other hand, the techniques used to handle most statistical housekeeping chores are usually very straightforward. They involve basic arithmetical and statistical functions: addition, division, multiplication, subtraction, interpolation, extrapolation, regression, as well as incorporating better (benchmark) data to revise or correct previously published estimates.

Interpolation and Extrapolation

Let's walk through two of these statistical techniques—interpolation and extrapolation—frequently used statistical methods to fill or close data gaps. To interpolate—to estimate missing values between two known values or known points in time on the basis of existing data—you need two years of hard data, a beginning and an end point, to serve as benchmarks. **Hard data** are data that are available and are statistically and conceptually consistent with the activity being measured. **Soft data** are partial or related data that can be made statistically and conceptually consistent with an activity.

If you really want to know the details, this is how it works for **interpolation**. Let's suppose that you want to estimate manufacturing payrolls for each year from 1980 through 1990, but you only have manufacturing payroll data for 1980 and 1990. Let's then suppose that you have hard manufacturing employment data for all the years 1980 through 1990. Also, you believe, based on past trends, that manufacturing payrolls will increase or decrease in proportion to the increase or decrease in reported manufacturing employment. In other words, you believe that there is a direct relationship between the movement in these two data series over time.

The first step is to compute the ratio of reported manufacturing payrolls to manufacturing employment in 1980 and 1990. The next step is to accept the convention that the ratios you computed for 1980 and 1990 changed by even amounts for the missing years. The final step is to apply these interpolated (or inserted) ratios to the reported manufacturing employment data to produce an interpolated series of manufacturing payroll estimates for the years 1981 through 1989. In this manufacturing example, you have hard or actual payroll data for the years 1980 and 1990 and soft or estimated payroll data for 1981 through 1989. The interpolated (or inserted) manufacturing payroll data for the intervening years, based on the manufacturing employment data, produces an estimate that approximates the actual manufacturing payroll data.

The procedure for extending or projecting data on the basis of existing data is called **extrapolation**. To extrapolate you need at least one year of hard data to serve as a benchmark. If you have manufacturing payroll data for 1990 and manufacturing employment data for 1990, 1991 and 1992, and want to estimate payrolls for 1991 and 1992, you use the relationship between manufacturing payrolls and employment in 1990 and assume the 1990 ratio would be the same for 1991 and 1992. You then apply the 1990 ratio to the 1991 and 1992 employment data to get a reasonable approximation or estimate of manufacturing payrolls for 1991 and 1992. But you can't be sure of the validity of the extrapolated results until you have either hard manufacturing payroll data for 1991 and 1992 or manufacturing payroll data for a later year.

For both interpolation and extrapolation, the reliability of the data depends on the validity of the assumptions about the relationship between the data sets.

VII. The Best Business-Related Data Publications

VII. The Best Business-Related Data Publications

Economic and Financial Data Publications

The three most important business and financial data publications are The Monthly Labor Review, The Survey of Current Business and the Federal Reserve Bulletin. Like the legendary multi-purpose Swiss army knife, these three government reports meet just about every business-related data need.

The Monthly Labor Review

The Monthly Labor Review (MLR) is a publication of the U.S. Department of Labor's Bureau of Labor Statistics. This is a monthly publication containing articles and reports on employment, prices, wages, productivity, job safety and a wide range of labor-related topics.

What makes the MLR especially valuable to data users, in addition to such regular features as a review of developments in industrial relations and book reviews, is that it includes a statistical section of 51 tables of current labor-related data.

The Survey of Current Business

The Survey of Current Business (SCB) is a monthly publication of the U.S. Department of Commerce's Bureau of Economic Analysis. It contains estimates and analyses of U.S. economic activity, including a section called "Business Situation"—a review of current economic developments—as well as regular and special articles pertaining to national, regional and international economic accounts and related topics.

What makes the SCB particularly useful is that it contains two statistical sections that include an array of economic data from various public and private sources:

> The Business Cycle Indicators section consists of 250 tables and 130 charts that are widely used in analyzing current business cycle developments.

> The Current Business Statistics section consists of tables for over 1,900 data series covering general business activities and specific industries.

The Federal Reserve Bulletin

The Federal Reserve Bulletin is a monthly publication of the Board of Governors of the Federal Reserve System. It contains articles and data on financial and business activities.

The Domestic Financial Statistics section includes: money, stock and bank credit; policy instruments, federal reserve banks; monetary and credit aggregates; commercial banking institutions, financial markets, federal finance; securities markets and corporate finance; real estate; consumer installment credit; and flow of funds.

The International Statistics section includes: summary statistics on U.S. international transactions; U.S. foreign trade; U.S. reserve assets; and other related data reported by banks and nonbanking business enterprises in the U.S.

The Domestic Nonfinancial Statistics section includes: selected measures of nonfinancial business activity, labor force, employment and unemployment; output, capacity and capacity utilization; industrial production; housing and construction; consumer and producer prices; gross national product and income; and personal income and savings.

Copies of the MLR and SCB can be ordered directly from BLS or BEA, from the Government Printing Office (GPO), and through the National Technical Information Service (NTIS).

To place subscriptions to the MLR or SCB through the Government Printing Office write to: Superintendent of Documents, Government Printing Office, 710 North Capitol Street, N.W., Washington, DC 20401, or call the GPO at **(202) 783-3238**.

For the Federal Reserve Bulletin write to: Publications Services, Mail Stop 138, Board of Governors of the Federal Reserve System, Washington, DC 20551, or call **(202) 452-3244** or **(202) 452-3245**.

To place subscriptions through NTIS write: U.S. Department of Commerce, National Technical Information Service, 5285 Port Royal Road, Springfield, VA 22161, or call **(703) 487-4650**.

VII. The Best Business-Related Data Publications

National and International Statistical Guides

National Statistics

The American Statistics Index (ASI), a product of the Congressional Information Service, is the best guide and index, not only to economic data publications, but to all the statistical publications of the U.S. government. The American Statistics Index:

> **1. Identifies** the statistical data published by all branches and agencies of the federal government.
>
> **2. Catalogs** the publications in which these data appear, providing full bibliographic information about each publication.
>
> **3. Announces** new publications as they appear.
>
> **4. Describes** the contents of these publications fully.
>
> **5. Indexes** this information in full subject detail; and
>
> **6. Micropublishes** virtually all the publications covered in ASI to provide reliable access to economic and other data.

ASI includes all federal publications that contain primary data of research value or secondary data collected on a special subject. It also has special studies and analysis or statistics related materials. All types of publications are included—periodicals, special one-time reports, items within a large continuing report series, and annual or biennial reports.

International Statistics

The Congressional Information Service began publishing the **Index of International Statistics (IIS)** in 1983. The IIS is a comprehensive monthly index and abstracting service covering the statistical publications of international intergovernmental organizations, including the United Nations (UN), Organization for Economic Cooperation and Development (OECD), European Economic Community (EEC), Organization of American States (OAS), and approximately 40 other organizations.

Non-Governmental Statistics

The Congressional Information Service began publishing the **Statistical Reference Index** (**SRI**) in 1980. It is a monthly abstract and index publication covering statistical reports from U.S. sources other than the federal government. These sources include trade, professional, and other nonprofit associations, business organizations, commercial publishers, independent research centers, state government agencies, and university research centers. SRI has selected from these sources a cross-section of documents presenting basic national and state data on business, industry, finance, economic and social conditions, the environment and population.

Educational Data Publications

One of the most important factors affecting the competitiveness of the United States is the educational level and quality of the U.S. labor force. Educational data can forecast the future quality of the labor force.

The primary source for educational statistics is the National Center for Educational Statistics (NCES). The Center gathers and publishes information on the status and progress of education in the U.S. The two most useful publications of the Center in terms of content and scope are **The Condition of Education** and the **Digest of Education Statistics**.

The Condition of Education includes 50 to 60 indicators—key data that measure the health of education, monitor important developments, and show trends in major aspects of education. The data are published in two volumes. The first volume covers elementary and secondary education and the second volume covers postsecondary education. Each volume includes the text, tables, and charts for each indicator plus the technical supporting data, supplemental information, and data sources. Although indicators may be simple statistics, more often they are analyses—examining relationships; showing changes over time; comparing or contrasting subpopulations, regions, and states. The NCES emphasizes that the data used for these indicators are the most valid and representative education statistics currently available in the U.S.

The **Digest of Education Statistics** includes more than 350 statistical tables, plus figures and appendices. These indicators represent a consensus of professional judgment on the most significant national measures of the condition and progress of education currently based on available current and valid data. New indicators for elementary and secondary education added to the **Digest of Education Statistics** include: high school dropout rates; course-taking patterns of high school students; the proportion of high school students who work while attending school; and eighth graders' attitudes about school climate. New indicators for postsecondary education include: the college enrollment rate for recent high school graduates; tuition charges as a fraction of income of families with children; proportion of young adults holding jobs by years of schooling completed; and the distribution of college students by parents' education and income.

The main difference between the **Digest of Education Statistics** and **The Condition of Education** is that the **Digest** covers many more topics, but with much less interpretation. To qualify for inclusion in the **Digest**, material must only be nationwide in scope and of current interest and value." **The Condition of Education** measures the size of the educational system and attempts to assess how well the system performs.

Part One: The DATAPRIMER

To obtain the **Digest of Education Statistics** or **The Condition of Education** write to: Superintendent of Documents, U.S. Government Printing Office, 710 North Capitol Street, N.W., Washington, DC 20401, or call **(202) 275-2091**.

Source: U.S. Department of Education, National Center for Education Statistics

VII. The Best Business-Related Data Publications

The Best Summary of U.S. Statistics

The **Statistical Abstract of the United States**, published since 1878, is the standard summary of statistics on the social, political, and economic organization of the United States. It is designed to serve as a convenient volume for statistical reference and as a guide to other statistical publications and sources.

It includes a selection of data from many statistical publications, both governmental and private. Publications cited as sources usually contain additional statistical detail and more comprehensive discussions of definitions and concepts than can be included in the **Statistical Abstract**. Data not available in publications issued by the contributing agency, but obtained from unpublished records are identified in the source notes as "unpublished data." Information on the subjects covered in tables so noted may generally be obtained from the source.

Although the emphasis in the **Statistical Abstract** is on national data, many tables present data for regions and individual states and to a lesser extent for metropolitan areas and cities. Additional information for states, cities, counties, metropolitan areas, and other small units, as well as more historical data, are available in various supplements to the **Abstract**.

The **Statistical Abstract** now includes data on death from AIDS, registered nurses, school enrollment projections by state, pet ownership, finances of Political Action Committees, child care, employment projections, cost of living indexes, the financial condition of insured commercial banks, employment drug testing, mergers and acquisitions, airline on-time arrivals and departures, and corporate farming.

The Guide to Tabular Presentation in the **Statistical Abstract** has been expanded to include more explanations on how to interpret the unit indicators presented in the tables.

Cost and ordering information can be obtained from the U.S. Bureau of the Census, Customer Services, Washington, DC 20233, or by calling **(301) 763-4100**.

Government Printing Office Bookstores

Government publications can be ordered from the U.S. Government Printing Office bookstores listed below:

Alabama
O'Neill Building
2021 3rd Avenue North
Birmingham, AL 35203
(205) 731-1056

California
ARCO Plaza, C-Level
505 South Flower Street
Los Angeles, CA 90071
(213) 239-9844

California
Federal Building, Room 1023
450 Golden Gate Avenue
San Francisco, CA 94102
(415) 252-5334

Colorado
Federal Building, Room 117
1961 Stout Street
Denver, CO 80294
(303) 844-3964

Colorado
World Savings Building
720 North Main Street
Pueblo, CO 81003
(719) 544-3142

District of Columbia
Farragut West
1510 H Street, N.W.
Washington, DC 20005
(202) 653-5075

District of Columbia and Vicinity
Government Printing Office
710 North Capital Street, N.W.
Washington, DC 20401
(202) 275-2091

Florida
Federal Building, Room 158
400 West Bay Street
Jacksonville, FL 32202
(904) 353-0567

Georgia
Federal Building, Room 100
275 Peachtree Street, NE
P.O. Box 56445
Atlanta, GA 30343
(404) 331-6947

Illinois
Federal Building, Room 1365
219 South Dearborn Street
Chicago, IL 60604
(312) 353-5133

VII. The Best Business-Related Data Publications

Maryland
Warehouse Outlet
8660 Cherry Lane
Laurel, MD 20707
(301) 953-7974

Massachusetts
Thomas P. O'Neill Federal Building
10 Causeway Street, Room 179
Boston, MA 02222
(617) 720-4180

Michigan
Federal Building, Suite 160
477 Michigan Avenue
Detroit, MI 48226
(313) 226-7816

Missouri
120 Bannister Mall
5600 East Bannister Road
Kansas City, MO 64137
(816) 765-2256

New York
Federal Building, Room 110
26 Federal Plaza
New York, NY 10278
(212) 264-3825

Ohio
Federal Building, Room 1653
1240 East 9th Street
Cleveland, OH 44199
(216) 522-4922

Ohio
Federal Building, Room 207
200 North High Street
Columbus, OH 31215
(614) 469-6956

Oregon
1305 SW First Avenue
Portland, OR 97201
(503) 221-6217

Pennsylvania
Robert Morris Building
100 North 17th Street
Philadelphia, PA 19103
(215) 597-0677

Pennsylvania
Federal Building, Room 118
1000 Liberty Avenue
Pittsburgh, PA 15222
(412) 644-2721

Texas
Federal Building, Room 1050
1100 Commerce Street
Dallas, TX 75242
(214) 767-0076

Texas
Texas Crude Building
801 Travis Street
Houston, TX 77002
(713) 228-1187

Washington
Federal Building, Room 194
915 Second Avenue
Seattle, WA 98174
(206) 442-4270

Wisconsin
Federal Building, Room 190
517 East Wisconsin Avenue
Milwaukee, WI 53202
(414) 291-1304

VIII. Jargon

Terms, Concepts and Measures

In any field, jargon separates the layman from the professional. So that you can bridge the jargon gap with regard to economic data, we have compiled a list of terms and concepts that will help you understand most of the terminology you might encounter when using economic data. These listings of buzz-words and lingo—the professional jargon of the data merchant's trade—will give you a quick grasp of essential statistical terms and measures.

In putting together this glossary of definitions and usages, we have used several sources, but we have drawn heavily on one of the bibles of the trade—the <u>Dictionary of Economic and Statistical Terms</u> published by the U.S. Department of Commerce.

In doing this, we have followed the advice of the late William H. Charterner, a past president of the National Association of Business Economists, who emphasized that an understanding of today's economic issues is important for our survival. He believed that every effort had to be made to give professionals who are not economists a clear understanding of the meanings and uses of economic terms and measures. We share that belief and prepared this chapter to serve as a reference to data users.

VIII. Jargon: Terms, Concepts and Measures

Commonly Used Economic and Statistical Terms

Administrative Record Data

Administrative record data are the by-product of the administration of federal, state and local tax programs, social security programs, unemployment insurance programs and other federal, state and local programs.

Annual Rates

Annual rates put values for one or more months at their annual equivalent. They are used to change quarterly and monthly data into yearly data to compare values for time periods of different lengths. For example, if 2 million cars were sold in a quarter, the annual rate of sales would be 8 million cars.

Benchmark Data

Benchmark data are comprehensive data (compiled at infrequent intervals) used as a basis for developing and adjusting interim estimates. Many of the monthly and quarterly estimates of economic data published by the government's Data Factories make use of decennial, quinquennial and annual censuses and surveys for this purpose.

Benefit-Cost Analysis

Benefit-cost analysis is an analysis or comparison of benefits in contrast to the costs of any particular action.

Business Cycle (Coincident) Indicators

The **coincident indicators** are a set of economic data that provide an analytical basis for determining the dates when the overall economy actually reaches peaks and troughs in economic activity. They are among the most important indicators of current business activity available.

Business Cycle (Composite Indexes) Indicators

Composite indexes combine selected cyclical indicators (leading, coincident, lagging) into weighted representative indexes. Data that lead the business cycle are combined into one index, those that roughly coincide with the cycle into another, those that lag into a third.

Part One: The DATAPRIMER

Business Cycle (Lagging) Indicators

The **lagging indicators** are a set of economic data which usually reaches business cycle peaks and troughs "after" changes in general economic activity.

Business Cycle (Leading) Indicators

The **leading indicators** are a set of economic data that provide clues to shifts in business activity, usually several months in advance of business activity. Business cycle highs and lows lead or precede economic activity.

Business Sector Productivity Measure

Business sector productivity is the broadest and most widely used measure of productivity. Business sector productivity measures the value-added in the business sector per hour worked. The business sector makes up roughly 75 percent of GDP. It excludes output from the rest-of-the-world, the government, output from paid employees of household help, nonprofit institutions, the rental value of owner-occupied homes, and the statistical discrepancy in computing the national income and product accounts.

Constant Dollar Estimates

Constant dollar estimates are produced by removing the effects of price changes from statistical series reported in dollar terms. Constant dollar series are derived by dividing current dollar estimates by appropriate price indexes. The result is prices that are the same as in the base year—as if the dollar had constant purchasing power.

Construction Industries

Construction industries are establishments engaged primarily in constructing new homes and other buildings—heavy construction like highways and special trades (plumbing and electrical work).

Consumer Price Index (CPI)

The **CPI** is a measure of the average change in prices paid by urban consumers for a fixed market basket of goods and services. It is calculated monthly for two population groups, one consisting only of wage earners and clerical workers and the other all urban families. The all urban index (CPI-U) is representative of the buying habits of about 80 percent of the U.S. population.

VIII. Jargon: Terms, Concepts and Measures

Corporate Profits

Corporate profits from current production reflect replacement costs for inventories and capital used. In contrast, "book profits" generally do not reflect expenses related to current production. For example, if an auto dealer purchases an automobile for $10,000 and sells it for $13,000, his "book profit" would be $3,000. If at the time he sold the automobile, it would cost him $11,000 to replace it in inventory, his "corporate profit" would be $2,000.

Current Population Survey (CPS)

The **CPS** is the major source for data on the labor force, employment, unemployment, and persons not in the labor force. It is classified by a variety of demographic, social and economic characteristics. It is a monthly survey by the Census Bureau of approximately 60,000 occupied households using a scientifically selected sample of households that are representative of the civilian, noninstitutional population of the United States.

Data Series

A **data series** is an economic measure or item covering a specific time span.

Demand

Demand is what one or all potential purchasers are willing to buy at certain prices. In contrast to effective demand, which measures the quantities that potential purchasers are able to buy.

Diffusion Index

A **diffusion index** is a statistical measure of the overall behavior of a group of economic time series. It indicates the percentage of series expanding in a selected group. If one-half of the series rises over a given time span, the diffusion index value equals 50. As an analytical measure, the diffusion index is helpful in indicating the spread of economic movements from one industry to another or from one economic process to another.

Disposable Personal Income

Disposable personal income is the income remaining to persons after personal tax and non-tax payments. Personal taxes include state and federal income taxes as well as personal property taxes, estate and gift taxes. Non-tax payments include fines and fees for services like education and hospitals.

Part One: The DATAPRIMER

Econometric Model

An **econometric model** is a set of related equations used to analyze economic data. Such models are devised in order to depict the essential quantitative relationships that determine the behavior of output, income, employment and prices. Econometric models are used for forecasting or estimating the likely quantitative impact of alternative assumptions.

Economic Time Series

An **economic time series** is a set of quantitative data collected over regular time intervals (weekly, monthly, quarterly, annually). It measures some aspect of economic activity.

Effective Demand

Effective demand is the desire for goods and services as measured by the quantity which one or all potential purchasers are ready and able to buy at prevailing prices.

Employment Cost Index (ECI)

The **ECI** measures the rate of change in employee compensation including wages, salaries, and employer cost for employee benefits. It covers all establishments and occupations in both the private nonfarm and public sectors. It measures the change in the cost of employing a fixed set of labor inputs. The survey is published quarterly. The ECI enables data users to compare rates of change in occupational, industrial, geographic, union coverage and ownership.

Exports

Exports measure the total physical movement of merchandise from the United States to foreign countries, whether such merchandise is exported from within U.S. Customs territory, from a U.S. Customs bonded warehouse or a U.S. Foreign Trade Zone.

Exports (Domestic)

Exports of domestic merchandise include commodities which are grown, produced or manufactured in the United States as well as commodities of foreign origin which have been changed in the United States (for example, in U.S. Foreign Trade Zones) from the form in which they were imported, or which have been enhanced in value by further manufacture in the United States.

VIII. Jargon: Terms, Concepts and Measures

Exports (Foreign or Re-Exports) — **Exports of foreign merchandise (re-exports)** consist of commodities of foreign origin which have entered the United States for consumption or placement into custom bonded warehouses of U.S. Foreign Trade Zones and when exported are in substantially the same condition as when they were imported.

Extrapolation — **Extrapolation** is a statistical method used to extend or project data on the basis of existing data.

Finance Industries — **Finance industries** are establishments operating depository institutions, nondepository credit institutions, holding (not predominantly operating) companies, other investment companies, brokers and dealers in securities and commodity contracts, and security and commodity exchanges.

Fixed-Weighted Price Index — The **fixed-weighted price index** is an index derived by summing the price indexes used to compute constant-dollar GDP, weighted by the corresponding components' portion of output in the base year. This index measures price changes for the market basket that comprised GDP in the base year. Because the weights are always the same, changes in the fixed-weighted price index reflect pure price movements.

Foreign Trade Zones — **Foreign trade zones** are enclosed areas, operated as public utilities, under the control of the U.S. Customs with facilities for handling, storing, manipulating, manufacturing and exhibiting goods.

Gross — **Gross** is an overall total exclusive of deductions.

Part One: The DATAPRIMER

Gross Domestic Product (GDP) **Gross domestic product** refers to the value of production that takes place within the United States. (Gross national product refers to the value of production that is attributed to labor and property supplied by the residents of the United States regardless of whether or not the production takes place within the United States.) For example, if an automobile factory located in Detroit is completely owned by a foreign company and all employees of the factory are residents of Canada, the automobiles produced by the factory would be included in gross domestic product, but would not be included in gross national product.

Gross National Product (GNP) **Gross national product** refers to the value of production that is attributed to labor and property supplied by the residents of the United States regardless of whether or not the production takes place within the United States. It is a gross measure because no deduction is made to reflect the wearing out of machinery and other capital assets used in production. It excludes the value of intermediate products—goods and services for resale or for further processing.

Implicit Price Deflator **Implicit price deflators** are the byproduct of the preparation of current- and constant-dollar measures. For example, the current-dollar GDP components are estimated and deflated by appropriate price indexes to determine constant-dollar GDP components. These components are then summed to current- and constant-dollar GDP. The implicit price deflator is derived by dividing current-dollar GDP by constant-dollar GDP (and multiplying by 100). Changes in the deflator reflect changes both in prices and in the composition of GDP. So, strictly speaking it not a measure of pure price movements. (See fixed-weighted price index.)

Imports **Imports** of merchandise include commodities of foreign origin as well as goods of domestic origin returned to the United States with no change in condition or after having been processed and/or assembled in other countries.

VIII. Jargon: Terms, Concepts and Measures

Imputations

Imputations are estimates which make possible the measurement of certain types of income that do not take measurable monetary form. The four major imputations made in the national income and product accounts are for wages and salaries paid in kind (food, clothing, lodging); rental value of owner-occupied houses; food and fuel produced and consumed on farms; and interest payments by financial intermediaries.

Income

Income represents, for each person 14 years old and over, the amount of money income received in the previous year from each of the following sources: money wages and salaries; net income from nonfarm self-employment; net income from farm self-employment; social security, veterans' payments, or other government or private pensions; interest (bonds or savings), dividends, and income from annuities, estates, or trusts; net income from boarders or lodgers, or from renting property to others; all other sources such as unemployment benefits, public assistance, alimony, etc. The amount of income received represents income before deductions for personal taxes.

Index Numbers

Index numbers are a measure of relative value compared to a base figure. Index numbers possess a number of advantages over the raw data from which they are derived. First, they facilitate analysis by their simplicity. Second, they are a more useful basis for comparison of changes in data originally expressed in dissimilar units. Third, they permit comparisons over time with a common starting point, the index base period.

Industry Productivity

Industry productivity measures gross output (sales) produced per unit of labor for each major industry.

Input-Output Tables

Input-output tables depict the way the industries of a nation interact. They show, for each industry, the amount of its output that goes to each other industry as raw materials or semifinished products, as well as the amount that goes to the final markets of the economy. They also show, for each industry, its consumption of the products of other industries, as well as its contribution to the production process in the form of value added. These tables permit the tracing of the industrial repercussions, direct and indirect, of changes in consumer demand, in demand for investment goods, exports, and government procurement.

Insurance Business

Insurance business includes carriers of all types of insurance, insurance agents and brokers.

Interpolation

Interpolation is a statistical method used to estimate missing values between two known values or known points in time, usually on the basis of existing related data.

Inventory Valuation Adjustment and Profits

The **inventory valuation adjustment** is the difference between the purchase cost of goods in inventory and their replacement cost at the time the goods are removed from inventory and sold. For example, if an automobile in inventory cost $10,000 when it was purchased and would cost $12,000 to replace at the time it was sold, the inventory valuation adjustment would be minus $2,000. This adjustment is required to prevent overstatement or understatement of earned profits in periods of changing prices.

Manufacturing Industries

Manufacturing industries are those industries engaged in the mechanical or chemical transformation of materials or substances into new products.

Mean

The **mean** is the arithmetic average of a set of values. It is derived by dividing the sum of numerical items by the number of items.

VIII. Jargon: Terms, Concepts and Measures

Median

The **median** is the middle value in a set of data arranged in order of increasing or decreasing magnitude. The median is often used in describing the typical case. For example, to say that the median family income in the United States in 1991 was $28,000 means that half of all families had incomes larger than that value and the other half had less.

Merchandise Trade Balance

The **merchandise trade balance** represents the difference between U.S. exports and U.S. imports. This balance corresponds to a measurement of the international payments or credit flows resulting from the physical movement of goods between the U.S. and foreign countries.

Merchandise Trade Data

Merchandise trade data reflect the flow of merchandise, but not intangibles like services and financial commitments. The trade figures trace commodity movements out of and into the U.S. Customs jurisdiction—Puerto Rico, the Virgin Islands, the 50 states and the District of Columbia. Both import and export data exclude merchandise shipped in transit through the United States from one foreign country to another. Exports include intra-company shipments and also foreign aid, military sales, and agricultural assistance commodities. They exclude certain shipments made by the federal government.

Mineral Industries

Mineral industries are establishments that extract minerals that occur naturally, whether in solid, liquid, or gaseous form, prepare them on site as required, and explore and develop mineral properties.

Multi-Factor Productivity

Multi-factor productivity measures changes in business sector output due to changes in capital, labor inputs, new technologies and other categories that can increase output without increasing the use of labor and capital. Multi-factor productivity is favored by most economists because it does not ascribe all changes in output to one factor.

Part One: The DATAPRIMER

National Income

National income is the total earnings of labor and property from the production of goods and services. It is the income earned (not necessarily received) by all individuals and enterprises in the country for a specified period.

National Income and Product Accounts

The **national income and product accounts (NIPAs)**, the most widely used part of the national economic accounts, show the value and composition of the nation's output and the distribution of incomes. The accounts include: estimates of gross national product (GNP), GNP price measures, and the goods that make up the GNP.

Net

Net is the amount of value or quantity remaining after all charges, expenses, outlays, and losses have been deducted from the total or gross value or quantity.

Net Exports of Goods and Services

Net exports of goods and services is the balance of goods and services (excluding transfers under military grants) as reported in the U.S. balance-of-payments statistics. Total exports of goods and services are a part of the gross national product because they are produced by the national economy. Imports of foreign goods and services are included in the purchases of consumers, governments and businesses.

Net National Product

Gross and **net national product** differ by the value of the capital that is used in the production process. The value of the capital used is measured as depreciation. Gross national product is the value of production before the deduction of depreciation. Net national product is the value after depreciation is deducted from gross national product.

Nominal Dollar Estimates

Nominal dollar estimates are estimates expressed in current dollars. No adjustment is made to the estimate to reflect differences due to changes in price and quality. (This is in contrast to real or constant dollar estimates which eliminate differences due to changes in price and quality.)

VIII. Jargon: Terms, Concepts and Measures

Nonsampling Error

Nonsampling errors can be attributed to a variety of sources resulting from the survey design: inability to obtain information about all cases in the sample; definitional difficulties; differences in the interpretation of questions; inability or unwillingness to provide correct information on the part of respondents; mistakes in recording or coding the data obtained; and other errors of collection, response, processing, coverage, and estimation for missing data.

Personal Consumption Expenditures

Personal consumption expenditures reflect the market value of goods and services purchased by individuals and nonprofit institutions or acquired by them as income in kind. Purchases are recorded at cost to consumers (including excise or sales taxes) at the time of purchase, whether by cash or credit. The rental value of owner-occupied dwellings is included, but not the purchase of dwellings.

Personal Income

Personal income is the income received by all individuals in the economy from all sources. It is made up of wages and salary disbursements, proprietors' income, rental income of persons, dividends, personal interest income, and transfer payments (less personal contributions for social insurance).

Personal Outlays

Personal outlays is made up of personal consumption expenditures, interest paid by consumers, and personal transfer payments to foreigners. It represents the disbursements made by individuals of that portion of personal income available after the payment of personal taxes.

Personal Savings

Personal savings is the excess of personal income over the sum of personal outlays and tax and nontax payments. It consists of the current savings of individuals (including owners of unincorporated businesses), non-profit institutions and private health, welfare, and trust funds.

Private Households

Private households includes private households employing workers who serve on or about the premises in occupations usually considered as domestic service.

Producer Price Index

Producer price indexes reflect the price trends of a constant set of goods and services representing the total output of an industry. Industry indexes are based on the Standard Industrial Classification (SIC) system and provide comparability with a wide assortment of industry-based data for other economic phenomena, including productivity, production, employment, wages and earnings.

Productivity Measures

Productivity measures are constructed as the ratio of real gross product in a sector to the corresponding inputs in the sector. It shows changes in output per unit of input. Changes in the productivity of the business cycle typically show patterns which differ substantially from long-term movements. No single productivity measure can be regarded as best for all purposes.

Public Administration

Public administration includes the executive, legislative, judicial, administrative and regulatory activities of federal, state, and local governments and government establishments engaged in international affairs. Government owned and operated business establishments are classified according to the activity in which they are engaged.

Quarterly Financial Report of Manufacturing, Mining, and Trade Corporations

The **Quarterly Financial Report of Manufacturing, Mining, and Trade Corporations** provides current estimates of income, assets, liabilities, stockholders' equity, and related financial and operating ratios classified by industry and asset size. This is the only Census Bureau program that collects profit and loss information.

Real Estate Business

Real estate business includes owners, lessors, buyers, sellers, agents, and developers of real estate.

VIII. Jargon: Terms, Concepts and Measures

Regression Analysis

Regression analysis is a statistical method used to measure the relationship among two (or more) variables such as income and spending, costs and activity, etc. It is usually used by data makers when the data needed isn't available, but the value of a related variable is available or can be determined. The regression technique can be used to establish and assess the relationship between data needed (the dependent variable), and the data whose value will be used to estimate the missing data (the independent variable).

Regulatory Report Data

Regulatory report data are based on government regulatory reporting requirements imposed on selected industries, such as transportation, banking, and communications.

Retail Trade Establishments

Retail trade establishments are establishments engaged in selling merchandise for personal or household consumption and rendering services incidental to the sale of the goods. The difference between retail and wholesale trade is based on the type of customer—retailers sell primarily to individuals while wholesalers sell primarily to businesses and institutions.

Sample

A **sample** is a group of items selected from a larger group for the purpose of estimating the properties of the total group. Most of the current estimates of economic performance published by the Data Factories are based on sample data. The data obtained on the basis of a sample are usually accurate enough in their reflection of the total group performance to identify trends.

Sampling Error

Sampling error is the error arising from the *use* of a sample, rather than a census or a complete count to make an estimate.

Part One: The DATAPRIMER

Seasonal Adjustments

Seasonal adjustments are statistical modifications made to compensate for fluctuations in quarterly or monthly data due to circumstances like weather, holidays, and tax payment dates. These seasonal movements are often so strong that they distort the underlying changes in economic data and tend to mask the trends that may be developing. For purposes of economic analysis therefore, it is often desirable to remove these distortions. For example, sales in retail stores are usually heaviest on Friday and Saturday. Therefore, sales are likely to be greater in the months with five Fridays and/or Saturdays than in months with only four all other factors being equal. Trading-day adjustments identify and remove such calendar-related variations from a time series. There are also adjustments for the date on which important holidays occur. For instance, Christmas buying traditionally begins immediately after Thanksgiving Day. When Thanksgiving is early, retail sales for November are greater than when the holiday falls later in the month.

Seasonally Adjusted Annual Rates

A **seasonally adjusted annual rate** indicates that data have been adjusted for seasonal variation. The conversion of a monthly figure to an annual rate is accomplished by multiplying that figure by 12.

Services

Services are actions performed by a person or a group in accordance with the desires or needs of others.

Service Industries

Service industries are those establishments engaged in providing services for individuals, business, government establishments and other organizations.

Service Sector

The **service sector** is the part of the industrial sector of the economy consisting of the non-goods producing industries.

SIC

The **Standard Industrial Classification (SIC)** is the statistical classification standard underlying all establishment-based federal economic statistics classified by industry. The SIC classification covers the entire field of economic activities and defines industries in accordance with the composition and structure of the economy.

52

VIII. Jargon: Terms, Concepts and Measures

SIPP The **Survey of Income and Program Participation (SIPP)** provides microdata records for households from a longitudinal survey. Each household is interviewed at 4 month intervals for 32 months. The survey produces data on labor force activity on over 50 types of income as well as on participation in various cash and noncash benefit programs. Data for employed persons include the number of hours and weeks worked, earnings, and weeks without a job. Core data also cover postsecondary school attendance, public or subsidized rental housing, low-income energy assistance, and school breakfast and lunch program participation.

SITC The **Standard International Trade Classification (SITC)** is a statistical classification of the commodities imported and exported from the U.S. This classification is needed for purposes of economic analysis and to facilitate the international comparison of trade-by-commodity data. In 1988, the SITC adopted the harmonized commodity description and coding system to make U.S. commodity data more nearly comparable to data of other countries.

Statistical (National) Discrepancy The nation's output (gross domestic product) is calculated from two principal points of view: the income approach (how much is earned) and the product approach (how much is produced). Both methods yield estimates of gross domestic product (GDP). The **statistical discrepancy** is the amount by which these two estimates differ. This number cannot be used by itself as a measure of the accuracy of GDP. The two sets of data may not be completely independent.

Supply **Supply** is the quantity of a commodity offered for sale at a single specified price.

Technological Unemployment **Technological unemployment** is the displacement of workers by new labor-saving devices and technological changes in production.

Transfer Payments

Transfer payments are income flows representing a change in the distribution of national wealth, but are not compensation for a current contribution to the production process. Generally, they are paid in monetary form. They include payments by government and business to individuals and nonprofit institutions. Business transfers include bad debts, charitable contributions and contest prizes.

Transportation, Communications, Electric, Gas, and Sanitary Services

Transportation, communications, electric, gas, and sanitary service establishments provide the public, and business enterprises with passenger and freight transportation, communications services and electricity, gas, steam, water, and sanitary services.

Unemployment

Unemployment is the inability of the labor market to absorb all able bodied and qualified adults willing to work.

Unemployment (Total)

Total unemployment is the sum of those formerly employed in an industry and those who either entered the labor force for the first time, or re-entered the labor force after a period of separation. Seasonally unemployed, the unemployable and the transitionally unemployed are excluded from the estimates of total unemployment.

Wholesale Trade Businesses

Wholesale trade businesses are establishments engaged in selling merchandise to retailers—industrial, commercial, institutional, farm, or professional users or other wholesalers acting as agents or brokers in buying or selling merchandise to such persons or companies. Functions frequently performed by wholesale establishments include: maintaining inventories of goods; extending credit; physically assembling, sorting and grading goods in large lots; breaking bulk and redistributing smaller lots; delivery; and refrigeration.

IX. Using Statistical Tables

Take a quick look at the sample table on the next page before you read this section. Understanding what is included in this table is essential to using data effectively.

Most of the business-related data you will encounter are published in the form of statistical tables. A good statistical table should be easy to read and it should also contain enough information to make the data in the body of the table usable. A table will usually have (1) a table number to distinguish it from all other tables in a series, (2) a title that mentions the major data items in the table, (3) definitions in the headnotes and footnotes of any terms that are open to alternative interpretations, and (4) a reference to the sources of the data.

A good table title will tell you what the table is all about. The headnotes and footnotes contain the details that are too complex to include in the table title. The headnote is found at the top of the table under the title. It tells you that numbers are rounded to either thousands, millions or billions. If the data are in dollars, it may indicate that the data are either in current or constant dollars, or that the data are quarterly data that are seasonally adjusted at annual rates. Use caution. The table stubs and table headings (column and row headings) are summary in nature and may require footnotes to provide additional detail on the content of the item listed in the stub or heading.

If you want to know if the data in the table really describe the economic activity you are interested in, you have to read the fine print! As you can see from the sample table, table notes contain information needed to have a clear understanding of the content of the table. The footnotes and endnotes deserve a careful reading as well. Don't assume you know what a particular statistic is measuring until you read these notes.

IX. *Using Statistical Tables*

Sample Statistical Table

Table 3.12.—Government Transfer Payments to Persons
[Billions of dollars]

	Line	1987	1988	1989	1990
Government transfer payments to persons	1	**521.3**	**555.9**	**602.0**	**661.7**
Federal	2	**401.8**	**425.9**	**458.7**	**498.2**
Benefits from social insurance funds	3	353.0	373.4	402.1	436.4
Old-age, survivors, and disability insurance	4	201.0	213.9	227.4	244.1
Hospital and supplementary medical insurance	5	81.9	86.5	97.8	107.9
Unemployment insurance	6	14.6	13.6	14.5	18.1
State	7	14.2	13.1	14.1	17.7
Railroad employees	8	.1	.1	.1	.1
Federal employees	9	.3	.3	.3	.3
Special unemployment benefits	10				
Federal employee retirement	11	44.9	48.1	50.6	53.9
Civilian[1]	12	26.5	28.6	29.9	31.8
Military[2]	13	18.4	19.5	20.7	22.1
Railroad retirement	14	6.5	6.7	7.0	7.2
Pension benefit guaranty	15	.2	.3	.3	.3
Veterans life insurance	16	1.7	1.7	1.8	1.9
Workers' compensation	17	1.2	1.3	1.4	1.5
Military medical insurance[3]	18	1.1	1.3	1.3	1.5
Veterans benefits	19	14.8	15.0	15.4	15.8
Pension and disability	20	14.2	14.6	15.0	15.6
Readjustment	21	.6	.4	.4	.3
Other[4]	22				
Food stamp benefits	23	10.6	11.2	12.2	14.7
Black lung benefits	24	1.5	1.5	1.5	1.4
Supplemental security income	25	10.0	10.7	11.6	12.5
Direct relief	26				
Earned income credit	27	1.4	2.7	4.0	4.4
Other[5]	28	10.5	11.4	12.0	13.0
State and local	29	**119.6**	**130.0**	**143.3**	**163.5**
Benefits from social insurance funds	30	37.4	40.9	44.3	47.8
State and local employee retirement	31	31.2	34.1	36.6	39.2
Temporary disability insurance	32	1.6	1.8	1.9	2.2
Workers' compensation	33	4.6	5.1	5.8	6.4
Public assistance	34	75.6	82.0	91.2	106.9
Medical care	35	49.8	55.0	62.9	75.7
Aid to families with dependent children	36	16.7	17.3	18.0	19.8
Supplemental security income	37	2.9	3.1	3.4	3.8
General assistance	38	2.6	2.7	2.8	3.0
Energy assistance	39	1.7	1.7	1.4	1.6
Other[6]	40	1.9	2.3	2.7	3.0
Education	41	4.0	4.2	4.6	5.2
Employment and training	42	.9	.9	.9	.9
Other[7]	43	1.6	1.9	2.2	2.6

1. Consists of civil service, foreign service, Public Health Service officers, Tennessee Valley Authority, and several small retirement programs.
2. Includes the Coast Guard.
3. Consists of payments for medical services for dependents of active duty military personnel at nonmilitary facilities.
4. Consists of mustering out pay, terminal leave pay, and adjusted compensation benefits.
5. Consists largely of payments to nonprofit institutions, aid to students, payments for medical services for retired military personnel and their dependents at nonmilitary facilities.
6. Consists of emergency assistance and medical insurance premium payments paid on behalf of indigents.
7. Consists largely of foster care, veterans benefits, Alaska dividends, and crime victim payments.

Source: U.S. Department of Commerce, Bureau of Economic Analysis

X. Additional Readings on Data Issues

Chadwick, Healy, Guide to Statistical Materials Produced by Government and Associations in the U.S., U.S. Library of Congress No. 86-32722, 1985.

Center for Environmental Statistics Development Staff, EPA, A Guide to Selected National Environmental Statistics in the U.S. Government, April 1992.

Cohn, Victor, News & Numbers: A Guide to Reporting Statistical Claims and Controversies in Health and Other Fields, Iowa State University Press, 1989.

Eveleth, Max Jr., How the Private Sector Uses Data and Technology, Proceedings of the 1973 Executive Seminar on "Data Uses in the Private Sector," Sponsored by the Census Bureau, 1973.

Federal Reserve Bank of Richmond, Macro Economic Data: A User's Guide, 1990.

Frumkin, Norman, A Guide to Economic Indicators, Armonk, New York, M.E. Sharpe, 1990.

Hudson Group, Where to Find Business Information: A World Guide, A Hudson Group Book, 2nd Ed., Wiley, 1982.

Huff, Darrell and Geis, Irving, How to Lie with Statistics, W.W. Norton, Inc., 1982.

Kenessey, Zoltan E., Industrial Production and Capacity Utilization, "Recent Developments at the Federal Reserve: Part 1," Business Economics, National Association of Business Economists, October 1988.

Liesnes, Thelma, One Hundred Years of Economic Statistics: United Kingdom, United States, Australia, Canada, France, Germany, Italy, Japan and Sweden, New York; Facts on File, c1989.

Morsink, James H., Seasonal Adjustment, Macro Economic Data: A User's Guide, Federal Reserve Bank of Richmond, 1990.

National Association of Business Economists, Report of the Statistics Committee, February 1988.

Office of Technology Assessment, Statistical Needs for a Changing U.S. Economy, OTA-BP-E-58, Washington, DC, U.S. Government Printing Office, September 1989.

Tanur, Judith M., Statistics: A Guide to Business and Economics, The Holden-Day Series in Probability and Statistics, A collection of Essays, Holden-Day Inc., 1976.

U.S. Department of Commerce, Dictionary of Economic and Statistical Terms, 1975.

U.S. Government, Glossary of Nonsampling Error Terms: An Illustration of a Semantic Problem in Statistics, Available through NTIS Document Sales, PB86-211547/AS, 1986.

U.S. Government, Quality in Establishment Surveys, Available through NTIS Document Sales, PB86-232451/AS, 1986.

U.S. Government, Report on Statistical Uses of Administrative Records, Available through NTIS Document Sales, PB88-232921, 1985.

U.S. Office of Management and Budget, Report of the Working Group on the Quality of Economic Statistics (to the Economic Policy Council), Washington, DC, April 1987.

U.S. Office of Management and Budget, Statistical Programs of the United States Government, Fiscal Year 1990.

Wallman, Katherine K., Losing Count: The Federal Statistical System, Population Trends and Public Policy, Number 16, Population Reference Bureau Inc., September 1988.

Webb, Roy H. and Whelpley, William, Labor Market Data, Federal Reserve Bank of Richmond Economic Review, 1989.

Webb, Roy H. and Willemse, Rob, Macroeconomic Price Indexes, Federal Reserve Bank of Richmond Economic Review, 1989.

Webb, Roy H., National Income and Product Accounts, Federal Reserve Bank of Richmond Economic Review, 1986.

Young, Allan, Evaluation of GNP Estimates, Survey of Current Business, Volume 67 Number 8, Bureau of Economic Analysis, U.S. Department of Commerce, August 1987.

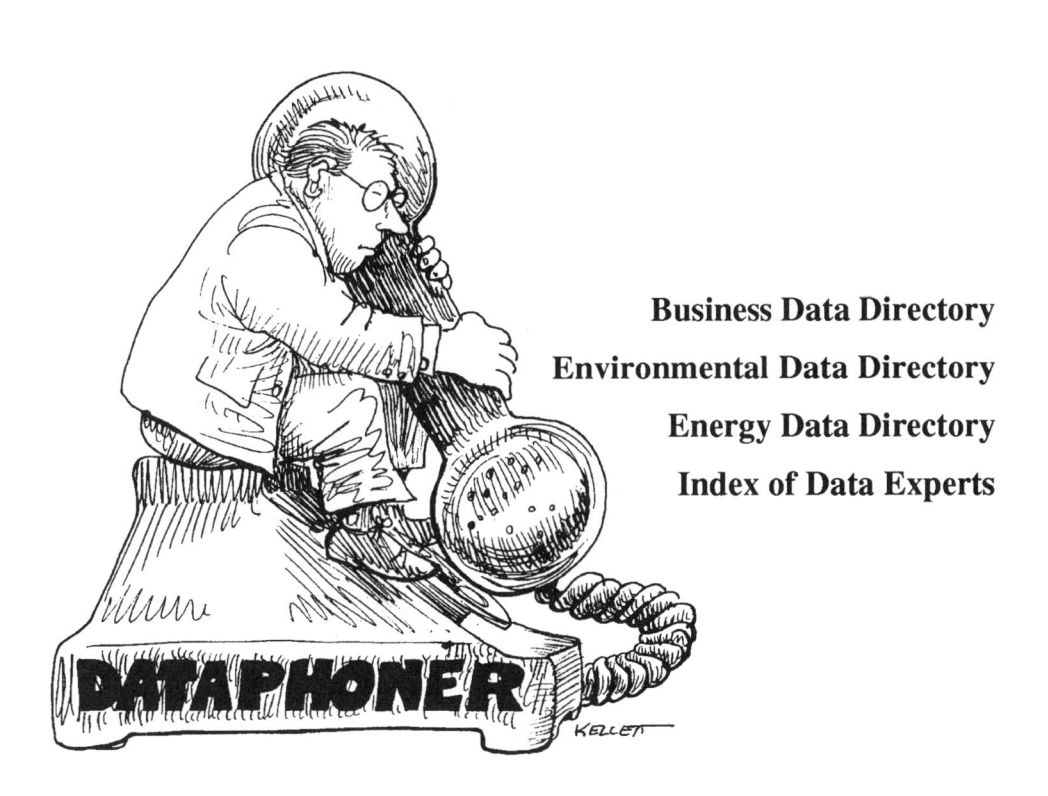

PART TWO
THE DATAPHONER

Business Data Directory

Environmental Data Directory

Energy Data Directory

Index of Data Experts

Using The Dataphoner

The **DATAPHONER** consists of three data directories:

The Business Data Directory
The Environmental Data Directory
The Energy Data Directory.

These three data directories contain over 2,500 data items and contacts. The **Business Data Directory** represents the core of the U.S. government's economic data network. It consists of three sections: Business Data Sources; State and Regional Data Centers; and Commercial Data Services. The State and Regional Data Centers section and the Commercial Data Services section are your guides to state and regional data—the names, telephone numbers and locations of more than 800 business, government and academic contacts in every one of the 50 states and the District of Columbia. These state data centers and commercial data service suppliers can provide state and local data, customize business-related data to meet your specific requirements, and answer questions on local data issues.

The **Environmental Data Directory** lists the contacts for the most frequently sought after business-relevant environmental statistics. The **Energy Data Directory** will put you in touch with the data managers of the business and environmentally significant energy data. It consists of two sections: Energy Data Sources and Uses of Energy Data.

Each directory contains the most authoritative contacts for current and practical data and information. Once you plug into this professional data network, you will get answers to your questions. And more than that, even if the people we have identified cannot help you, they can surely direct you to someone who can.

In the directories you may find the same subject or data expert listed several times. This is because in the data factories several professionals often share responsibility for the same subject while other professionals are often responsible for several subjects.

Part Two: The DATAPHONER

Keep in mind that there are two basic types of experts. First, there are the talking heads featured in the media, who use data selectively and are responsible for much of the current confusion about the economy, the environment and our energy problems. Fortunately for all of us there is another, much better informed group of experts—the economists, statisticians and specialists who work in the government's Data Factories. They are the people who take pride in their professional ability to work with data. The people listed in the **DATAPHONER** fall into this latter category of real data pros.

Tips on "Relating" to the Data Pros

Now you are ready to use the DATAPHONER. Find the appropriate subject and dial. Be polite and ask for the person you want. If the data contact you called is not available, **be persistent** and speak to someone in that office. **Don't be shy.** If the person you are speaking to can't help you, ask to talk to someone else.

If it appears that nobody is immediately available at the number you called, leave your name, your organization's name and telephone number. **Request that they return your call as soon as possible.** Call again in the afternoon or early the next morning.

If you are in a hurry contact one of the **Data Managers** listed under a separate heading in the Business Data Directory. They are the supervisors in the business-related Data Factories. These are busy people, but they usually can answer just about any question you might have. If they don't have the answer, they can refer you to someone who does have the answer.

Start your conversation with the data expert by telling him or her **exactly what you are trying to do** with the data. When it comes to recommending data for a specific purpose, the pros are usually the best judge about which data will best meet your needs.

Don't start your inquiry with a **shotgun request**. Be specific. Nothing will damage your credibility with the data pro more than saying, "I need **everything** you have on ...." Remember these are experts on specific areas of the economy and environment.

Remember that you will probably be calling this person again. **A "Thank you," goes a long way.** You might even send them a copy of your final report as a courtesy.

XI. Business Data Directory

Business Data Sources
State and Regional Data Centers
Commercial Data Services

XI. Business Data Directory

Section I. Business Data Sources

Business Data Sources, with 1,500 business data items, is your guide to the names and telephone numbers of the experts who produce and disseminate the nation's business data. This section gives you direct access to a broad range of primary business-related data. The items in the business data directory are organized into the major categories listed below so that you can quickly zero in on the information you want.

Major Categories

Agricultural Data
Banking and Financial Data
Business and Industry Data
Data Managers
Data Products and Other Information Services
Employment and Labor Force Data
Federal, State and Local Government Data
Geographic Concepts and Products
Housing Data
International and Foreign Trade Data
Personal Income and Related Data
Population and Other Demographic Data
Price and Price Index Data
Rural Development Data
Statistical Concepts and Methods
Unemployment Data
Working Conditions and Compensation Data

Organizational Index and Acronyms

BEA	Bureau of Economic Analysis, U.S. Department of Commerce
BLS	Bureau of Labor Statistics, U.S. Department of Labor
CENSUS	Bureau of the Census, U.S. Department of Commerce
DHHS	U.S. Department of Health and Human Services
EBRI	Employee Benefit Research Institute
ERS	Economic Research Service, U.S. Department of Agriculture
FRS	Federal Reserve System
HA	Hewitt Associates
HCBS	Health Care Benefits Survey
HCFA	Health Care Financing Administration, U.S. Department of Health and Human Services
HIAA	Health Insurance Association of America
NASS	National Agricultural Statistics Service, U.S. Department of Agriculture
NFIB	National Federation of Independent Businesses
NIAAA	National Institute of Alcohol Abuse and Alcoholism
OBA	Office of Business Analysis, U.S. Department of Commerce
PSRF	Profit Sharing Research Foundation
PWBA	Pension and Welfare Benefits Administration, U.S. Department of Labor
SBA	Small Business Administration
WB-DOL	Women's Bureau, U.S. Department of Labor
WLDF	Women's Legal Defense Fund

Agricultural Data

Subject	Source	Data Contact	Telephone
Africa	ERS	Kurtzig, Michael	202-219-0680
Alternative Crops	ERS	Glaser, Lewerne	202-219-0788
Aquaculture, Prices and Economic Data	ERS	Harvey, Dave	202-219-0890
Aquaculture, Production and Stock Data	NASS	Little, Robert	202-720-6147
Aquaculture, Production and Stock Data	NASS	Moore, Joel	202-720-3244
Asia, East	ERS	Coyle, William	202-219-0610
Asia, South	ERS	Landes, Rip	202-219-0664
Biotechnology	ERS	Reilly, John	202-219-0450
Canada	ERS	Simone, Mark	202-219-0610
Cash Receipts	ERS	Williams, Robert	202-219-0804
Cash Receipts	ERS	Dixon, Connie	202-219-0804
Cattle, Prices and Economic Data	ERS	Gustafson, Ron	202-219-1286
Cattle, Production and Stock Data	NASS	Shipler, Glenda	202-720-3040
China	ERS	Tuan, Francis	202-219-0626
Coffee and Tea Statistics	ERS	Gray, Fred	202-219-0888
Cold Storage	NASS	Lange, John	202-382-9185
Commodity Programs and Policies, World	ERS	Dixit, Praveen	202-219-0632
Commodity Programs and Policies	ERS	Harwood, Joy	202-219-0840
Commodity Programs and Policies	ERS	Hoff, Fred	202-219-0883
Commonwealth of Independent States (USSR)	ERS	Zeimetz, Kathryn	202-219-0624
Consumer Food Price Index	ERS	Dunham, Denis	202-219-0870
Consumer Price Index, Food	ERS	Parlett, Ralph	202-219-0870
Corporate Farms	ERS	Reimund, Donn	202-219-0522
Costs and Returns Data	ERS	Morehart, Mitch	202-219-0801
Costs and Returns Data	ERS	Dismukes, Robert	202-219-0801
Cotton, Prices and Other Economic Data	ERS	Skinner, Robert	202-219-0841
Cotton, Prices and Economic Data	ERS	Meyer, Leslie	202-219-0840
Cotton, Production and Stock Data	NASS	Latham, Roger	202-720-5944
Cotton, World Prices and Economic Data	ERS	Whitton, Carolyn	202-219-0824
Credit and Financial Markets	ERS	Sullivan, Pat	202-219-0719
Credit and Financial Markets, World Data	ERS	Baxter, Tim	202-219-0706
Credit and Financial Markets	ERS	Stam, Jerry	202-219-0892
Credit and Financial Markets	ERS	Ryan, Jim	202-219-0798
Crop Commodity Programs and Policies	ERS	Westcott, Paul	202-219-0840
Crop Production Costs	ERS	Dismukes, Robert	202-219-0801
Crop Statistics	Census	Jahnke, Donald	301-763-8567

Subject	Source	Data Contact	Telephone
Crops, Alternative	ERS	Glaser, Lewerne	202-219-0888
Crops Commodity Programs and Policies	ERS	Evans, Sam	202-219-0840
Dairy Biotechnology	ERS	Fallert, Richard	202-219-0710
Dairy Commodity Programs and Policies	ERS	Fallert, Richard	202-219-0710
Dairy Product Prices and Economic Data	ERS	Short, Sara	202-219-0769
Dairy Product Statistics	ERS	Mathews, Ken	202-219-0770
Dairy Product Statistics	NASS	Buckner, Dan	202-720-4448
Dairy Production Costs	ERS	Mathews, Ken	202-219-0770
Data Products	Census	Miller, Douglas	301-763-8561
Developing Economies	ERS	Mathia, Gene	202-219-0680
Dry Edible Beans	ERS	Lucier, Gary	202-219-0884
Dry Edible Beans	NASS	Budge, Arvin	202-720-4285
Dry Edible Beans	ERS	Greene, Catherine	202-219-0886
East Asia	ERS	Coyle, William	202-219-0610
Eastern Europe	ERS	Koopman, Robert	202-219-0621
Eastern Europe	ERS	Cochrane, Nancy	202-219-0621
Eastern Europe	ERS	Gray, Kenneth	202-219-0621
Economic Linkages to Agriculture	ERS	Edmondson, Bill	202-219-0785
Eggs, Production and Stock Data	NASS	Little, Robert	202-720-6147
Energy Statistics	ERS	Gill, Mohinder	202-219-0464
Export Commodity Programs and Policies	ERS	Smith, Mark	202-219-0821
Export Programs	ERS	Ackerman, Karen	202-219-0820
Exports	ERS	MacDonald, Steve	202-219-0822
Exports	ERS	Warden, Thomas	202-219-0822
Family Farms	ERS	Reimund, Donn	202-219-0522
Farm Economics	Census	Liefer, James	301-763-8566
Farm Household Income	ERS	Ahearn, Mary	202-219-0807
Farm Income	ERS	McElroy, Bob	202-219-0800
Farm Income	ERS	Strickland, Roger	202-219-0804
Farm Income	BEA	Smith, George	202-523-0821
Farm Labor	ERS	Whitener, Leslie	202-219-0932
Farm Labor	NASS	Kurtz, Tom	202-690-3228
Farm Labor Laws	ERS	Runyon, Jack	202-219-0932
Farm Labor Market	ERS	Whitener, Leslie	202-219-0932
Farm Land Ownership and Tenure	ERS	Wunderlich, Gene	202-219-0425
Farm Machinery	ERS	Vesterby, Marlow	202-219-0422
Farm Output	BEA	Smith, George	202-523-0821
Farm Output	ERS	Douvelis, George	202-219-0432
Farm Output, World	ERS	Urban, Francis	202-219-0717
Farm Prices, Parity Received	NASS	Buche, John	202-720-5446
Farm Prices, Parity Paid	NASS	Kleweno, Doug	202-720-4214

Agricultural Data XI. *Business Data Directory*

Subject	Source	Data Contact	Telephone
Farm Prices, Parity and Indexes	NASS	Milton, Bob	202-720-3570
Farm Product	BEA	Smith, George	202-523-0821
Farm Productivity	ERS	Douvelis, George	202-219-0432
Farm Real Estate Taxes	ERS	DeBraal, Peter	202-219-0425
Farm Structure	ERS	Reimund, Donn	202-219-0522
Farm Subsidies	ERS	Mabbs-Zeno, Carl	202-219-0631
Farm Subsidies	ERS	Nelson, Fred	202-219-0689
Farm Taxes	ERS	Durst, Ron	202-219-0896
Farm Wages	ERS	Whitener, Leslie	202-219-0932
Farm Wages	NASS	Kurtz, Tom	202-690-3228
Farms, Number of Farms	NASS	Ledbury, Dan	202-720-1790
Feed Grains, Prices and Economic Data	ERS	Tice, Thomas	202-219-0840
Feed Grains, Prices and Economic Data	ERS	Cole, James	202-219-0840
Feed Grains, Production and Stock Data	NASS	Van Lahr, Charles	202-720-7369
Feed Grains, World Prices and Economic Data	ERS	Riley, Peter	202-219-0824
Fertilizer	ERS	Taylor, Harold	202-219-0164
Fertilizer	ERS	Rives, Sam	202-720-2324
Finance and Trade Policy	ERS	Stallings, David	202-219-0688
Finance and Trade Policy	ERS	Sharples, Jerry	202-219-0791
Finance and Trade Policy	ERS	Baxter, Tim	202-219-0706
Finance and Trade Policy	ERS	Roningen, Vernon	202-219-0631
Finance and Trade Policy	ERS	Magiera, Steve	202-219-0633
Finances	ERS	Morehart, Mitch	202-219-0801
Finances	ERS	Hacklander, Duane	202-219-0798
Floriculture, Prices and Economic Data	ERS	Johnson, Doyle	202-219-0884
Floriculture, Production and Stock Data	NASS	Brewster, Jim	202-219-7688
Food Aid Programs	ERS	Suarez, Nydia	202-219-0821
Food Aid Statistics	ERS	Kurtzig, Michael	202-219-0680
Food Assistance and Nutrition	ERS	Smallwood, Dave	202-219-0864
Food Away from Home	ERS	Price, Charlene	202-219-0866
Food Consumption	ERS	Putnam, Judy	202-219-0870
Food Demand and Expenditures, World	ERS	Stallings, Dave	202-219-0708
Food Demand and Expenditures	ERS	Haidacher, Richard	202-219-0870
Food Demand and Expenditures	ERS	Blaylock, James	202-219-0862
Food Grains, World Prices and Economic Data	ERS	Schwartz, Sara	202-219-0824
Food Manufacturing	ERS	Gallo, Tony	202-219-0866
Food Manufacturing and Retailing	ERS	Handy, Charles	202-219-0866
Food Policy	ERS	Westcott, Paul	202-219-0840
Food Policy	ERS	Smallwood, Dave	202-219-0864
Food Policy	ERS	Meyers, Les	202-219-0860
Food Policy, World	ERS	Lynch, Loretta	202-219-0689

Subject	Source	Data Contact	Telephone
Food Prices and Consumer Price Index	ERS	Parlett, Ralph	202-219-0870
Food Prices and Consumer Price Index	ERS	Dunham, Denis	202-219-0870
Food Retailing	ERS	Kaufman, Phil	202-219-0866
Food Safety and Quality	ERS	Roberts, Tanya	202-219-0864
Food Wholesaling	ERS	Epps, Walter	202-219-0866
Foreign Agricultural Land Ownership	ERS	DeBraal, Peter	202-219-0425
Fruits and Tree Nuts, Production and Stock Data	NASS	Brewster, Jim	202-720-7688
Fruits and Tree Nuts, Production and Stock Data	NASS	Hintzman, Kevin	202-720-5412
Fruits and Tree Nuts, Prices and Economic Data	ERS	Bertelsen, Diane	202-219-0884
Fruits and Tree Nuts, Prices and Economic Data	ERS	Shields, Dennis	202-219-0884
Futures Markets, Agriculture	ERS	Heifner, Richard	202-219-0868
Futures Markets, Crops	ERS	Evans, Sam	202-219-0841
Futures Markets, Livestock	ERS	Nelson, Ken	202-219-0712
General Information	Census	Manning, Tom	301-763-1113
Guam, Agricultural Data	Census	Hoover, Kent	301-763-8564
Hay, Prices and Economic Data	ERS	Tice, Thomas	202-219-0840
Hay, Production and Stock Data	NASS	Eldridge, Herb	202-720-7621
History, Agriculture	ERS	Bowers, Douglas	202-219-0787
Hogs, Prices and Economic Data	ERS	Spinelli, Felix	202-219-0713
Hogs, Production and Stock Data	NASS	Fuchs, Doyle	202-720-3106
Honey, Price Data	NASS	Schuchardt, Rick	202-690-3236
Honey, Prices and Economic Data	ERS	Hoff, Fred	202-219-0883
Honey, Production Data	NASS	Kruchten, Tom	202-690-4870
Horticulture Statistics and Special Surveys	Census	Blackledge, John	301-763-8560
Imports	ERS	MacDonald, Steve	202-219-0822
Imports	ERS	Warden, Thomas	202-219-0822
Irrigation Statistics	Census	Blackledge, John	301-763-8560
Labor Market, Farm	ERS	Whitener, Leslie	202-219-0932
Land Ownership, Foreign, Agriculture	ERS	DeBraal, Peter	202-219-0425
Land Use	ERS	Daughtery, Arthur	202-219-0420
Latin America	ERS	Link, John	202-219-0660
Livestock Production Costs	ERS	Shapouri, H.	202-219-0770
Livestock Statistics	Census	Hutton, Linda	301-763-8569
Livestock, World Prices and Economic Data	ERS	Shagam, Shayle	202-219-0767
Livestock, World Prices and Economic Data	ERS	Bailey, Linda	202-219-1286
Macroeconomic Conditions	ERS	Monaco, Ralph	202-219-0782
Macroeconomic Conditions, World Data	ERS	Baxter, Tim	202-219-0706
Marketing Margins and Statistics	ERS	Elitzak, Howard	202-219-0870
Marketing Margins and Statistics	ERS	Handy, Charles	202-219-0866
Marketing Margins and Statistics	ERS	Dunham, Denis	202-219-0870
Middle East	ERS	Kurtzig, Michael	202-219-0680

Subject	Source	Data Contact	Telephone
Mink, Production and Stock Data	NASS	Kruchten, Tom	202-690-4870
Natural Resource Policy, World	ERS	Urban, Francis	202-219-0717
Natural Resource Policy	ERS	Ribaudo, Marc	202-219-0444
Natural Resource Policy	ERS	Heimlich, Ralph	202-219-0422
Natural Resource Policy	ERS	Osborn, Tim	202-219-0401
Northern Marianas	Census	Larson, Odell	301-763-8226
Nuts, Prices and Economic Data	ERS	Johnson, Doyle	202-219-0884
Nuts, Production and Stock Data	NASS	Hintzman, Kevin	202-720-5412
Organic Farming	ERS	Frerichs, Stephen	202-219-0401
Pacific Rim	ERS	Coyle, William	202-219-0610
Peanuts, Prices and Economic Data, World	ERS	McCormick, Ian	202-219-0840
Peanuts, Prices and Economic Data, World	ERS	Sanford, Scott	202-219-0840
Peanuts, Production and Stock Data	NASS	Latham, Roger	202-720-5944
Pesticides	ERS	Vandeman, Ann	202-219-0433
Pesticides	ERS	Rives, Sam	202-720-2324
Pesticides	ERS	Delvo, Herman	202-219-0456
Pesticides	ERS	Padgitt, Merritt	202-219-0433
Pesticides	ERS	Love, John	202-219-0886
Population	ERS	Swanson, Linda	202-219-0535
Population	ERS	Beale, Calvin	202-219-0535
Population, World Data	ERS	Urban, Francis	202-219-0705
Potatoes, Prices and Economic Data	ERS	Zepp, Glenn	202-219-0883
Potatoes, Prices and Economic Data	ERS	Lucier, Gary	202-219-0884
Potatoes, Production and Stock Data	NASS	Budge, Arvin	202-720-4285
Poultry, Prices and Economic Data	ERS	Christensen, Lee	202-219-0714
Poultry, Production and Stock Data	NASS	Moore, Joel	202-720-3244
Poultry, Production and Stock Data	NASS	Kruchten, Tom	202-690-4870
Poultry, Production and Stock Data	NASS	Little, Robert	202-720-6147
Poultry, Production and Stock Data	ERS	Perez, Agnes	202-219-0714
Poultry, World Prices and Economic Data	ERS	Witucki, Larry	202-219-0766
Price Spreads, Fruit and Vegetables	ERS	Parrow, Joan	202-219-0883
Price Spreads, Meat	ERS	Duewer, Larry	202-219-0712
Production Costs, Farms	NASS	Kleweno, Doug	202-720-4214
Productivity, Farm	ERS	Douvelis, George	202-219-0432
Productivity, Farm, World Data	ERS	Urban, Francis	202-219-0717
Proprietors' Income, Farm, Subnational Data	BEA	Zavrel, James	202-254-6638
Puerto Rico	Census	Hoover, Kent	301-763-8564
Rice, Prices and Economic Data	ERS	Levezey, Janet	202-219-0840
Rice, Prices and Economic Data	ERS	Young, Edwin	202-219-0840
Rice, Production and Stock Data	NASS	Owens, Marty	202-720-2157
Seeds	ERS	Gill, Mohinder	202-219-0464

Subject	Source	Data Contact	Telephone
Sheep, Prices and Economic Data	ERS	Stillman, Richard	202-219-0714
Sheep, Production and Stock Data	NASS	Simpson, Linda	202-720-3578
Soil Conservation	ERS	Osborn, Tim	202-219-0405
Soil Conservation	ERS	Magleby, Richard	202-219-0435
South Asia	ERS	Landes, Rip	202-219-0664
Soybeans, Prices and Economic Data	ERS	Hoskin, Roger	202-219-0840
Soybeans, Prices and Economic Data	ERS	McCormick, Ian	202-219-0840
Soybeans, Production and Stock Data	NASS	Vanderberry, Herb	202-720-9526
Sugar Commodity Programs and Policies	ERS	Lord, Ron	202-219-0888
Sugar, Prices and Economic Data	ERS	Buzzanell, Peter	202-219-0888
Sugar, Prices and Economic Data	ERS	Lord, Ron	202-219-0888
Sugar, Production and Stock Data	NASS	Eldridge, Herb	202-720-7621
Sunflowers, Prices and Economic Data	ERS	McCormick, Ian	202-219-0840
Sunflowers, Production and Stock Data	NASS	Vanderberry, Herb	202-720-9526
Sustainable Agriculture	ERS	Vandeman, Ann	202-219-0433
Sustainable Agriculture	ERS	Gajewski, Greg	202-219-0883
Sweeteners, Prices and Economic Data	ERS	Lord, Ron	202-219-0888
Sweeteners, Prices and Economic Data	ERS	Buzzanell, Peter	202-219-0888
Sweeteners, Production and Stock Data	NASS	Eldridge, Herb	202-720-7621
Sweeteners, Production Costs	ERS	Lord, Ron	202-219-0888
Tobacco Commodity Programs and Policies	ERS	Grise, Verner	202-219-0890
Tobacco, Prices and Economic Data	ERS	Grise, Verner	202-219-0890
Tobacco, Prices and Economic Data	ERS	Capehart, Tom	202-219-0890
Tobacco, Production and Stock Data	NASS	Eldridge, Herb	202-720-7621
Tobacco Production Costs	ERS	Clauson, Annette	202-219-0890
Trade and Finance Policy	ERS	Sharples, Jerry	202-219-0791
Trade and Finance Policy	ERS	Stallings, David	202-219-0688
Trade and Finance Policy	ERS	Magiera, Steve	202-219-0633
Trade and Finance Policy	ERS	Roningen, Vernon	202-219-0631
Trade and Finance Policy	ERS	Baxter, Tim	202-219-0706
Transportation, Agriculture	ERS	Hutchinson, T.Q.	202-219-0840
USSR, (Commonwealth of Independent States)	ERS	Zeimetz, Kathryn	202-219-0624
Vegetables, Fresh, Production and Stocks	NASS	Brewster, Jim	202-720-7688
Vegetables, Prices and Economic Data	ERS	Hamm, Shannon	202-219-0886
Vegetables, Prices and Economic Data	ERS	Lucier, Gary	202-219-0884
Vegetables, Prices and Economic Data	ERS	Greene, Catherine	202-219-0886
Vegetables, Processed, Production and Stocks	NASS	Budge, Arvin	202-720-4285
Vegetables, Production and Stock Data	NASS	Hintzman, Kevin	202-720-5412
Virgin Islands	Census	Hoover, Kent	301-763-8564
Water and Irrigation	ERS	Hostetler, John	202-219-0410
Water and Irrigation	ERS	Gollehon, Noel	202-219-0410

Subject	Source	Data Contact	Telephone
Water Quality	ERS	Rives, Sam	202-720-2324
Water Quality	ERS	Ribaudo, Marc	202-219-0444
Water Quality	ERS	Crutchfield, Steve	202-219-0444
Weather	ERS	Teigan, Lloyd	202-219-0705
Weather	NASS	Owens, Marty	202-720-2157
Western Europe	ERS	Coyle, William	202-219-0610
Wheat, Prices and Economic Data	ERS	Allen, Ed	202-219-0841
Wheat, Production and Stock Data	NASS	Siegenthaler, V.	202-720-8068
Wool and Mohair, Prices and Economic Data	ERS	Skinner, Robert	202-219-0841
Wool and Mohair, Prices and Economic Data	ERS	Lawler, John	202-219-0840
Wool, Production and Stock Data	NASS	Simpson, Linda	202-720-3578

Banking and Financial Data

Subject	Source	Data Contact	Telephone
Affiliates, Foreign	FRS	Martinson, M.	202-452-3640
Aggregate Reserves	FRS	Ribble, Leigh	202-452-2385
Agricultural Credit	FRS	Walraven, N.	202-452-2655
Agriculture	FRS	Rosine, John	202-452-2971
Allocation of System Open Market Account	FRS	Bettge, Paul	202-452-3174
Analytic Services, FR Bank Performance / Measures	FRS	Richards, Ms.	202-452-2705
Automobile Credit, Analysis	FRS	Luckett, Charles	202-452-2925
Automobile Credit, Data	FRS	Middleton, Millie	202-452-2924
Automobile Loans	FRS	Luckett, Charles	202-452-2925
Balance of Payments, U.S.	FRS	Morisse, Kathryn	202-452-3773
Balance Sheet for the U.S. Economy	FRS	Fogler, Elizabeth	202-452-3491
Balance Sheets for Federal Reserve Banks	FRS	Bettge, Paul	202-452-3174
Bank Acquisitions	FRS	O'Rourke, M.	202-452-3288
Bank Credit Cards	FRS	Bowman, Sharon	202-452-3667
Bank for International Settlements	FRS	Siegman, Charles	202-452-3308
Bank Holding Companies (BHCs), Domestic	FRS	Davenport, Ron	202-452-3623
Bank Holding Companies (BHCs), Acquisitions	FRS	Kline, Don	202-452-3421
Bank Loans	FRS	Farley, Dennis	202-452-3021
Bank Rates on Business Loans	FRS	McLaughlin, Mary	202-452-2259
Banking, International	FRS	Frankel, Allen	202-452-3578
Banking, International	FRS	Ryback, William	202-452-2722
Banking, International	FRS	Promisel, Larry J.	202-452-3533
Banking Markets	FRS	Rhoades, Stephen	202-452-3906
Banking Statistics System	FRS	Emerson, M.	202-452-2045
Banking Structure	FRS	Amel, Dean	202-452-2911
Banks, Miscellaneous Questions	FRS	Jemnings, Jack	202-452-3053
BHCs, Analysis of Mergers and Acquisitions	FRS	Burke, James	202-452-2612
Bills of Exchange	FRS	McDivitt, Patrick	202-452-3818
Bond Markets	FRS	Rea, John	202-452-3631
Bond Markets, Corporate	FRS	Crabbe, Leland	202-452-3022
Bond Markets, Municipal, Tax-Exempt	FRS	Turner, Chris	202-452-2983
Borrowers of Securities Credit (Margins)	FRS	Homer, Laura	202-452-2781
Broker Loans	FRS	Homer, Laura	202-452-2781
Business Finance	FRS	Rea, John	202-452-3744
Business Fixed Investment	FRS	Oliner, Stephen	202-452-3134
C&CA Computation of Reserve Requirement	FRS	Bethea, Martha	202-452-3181

Subject	Source	Data Contact	Telephone
Call Reports	FRS	Mitchell, Susanne	202-452-3684
Capacity Utilization	FRS	Raddock, Richard	202-452-3197
Capital Adequacy Guidelines	FRS	Cole, Roger	202-452-2618
Capital Flows, International	FRS	Stekler, Lois	202-452-3716
Capital Markets	FRS	Rea, John	202-452-3631
Certificates of Deposits, Banks	FRS	Reid, Brian	202-452-3589
Certificates of Deposits, Thrifts	FRS	Passmore, Wayne	202-452-6432
Chart Book Figures	FRS	Dykes, Ellen	202-452-3952
Commercial Paper, Interest Rates	FRS	McMillian, D.	202-452-2851
Commercial Paper, Outstanding	FRS	Post, Mitchell	202-452-2720
Competitive Analysis of Mergers and Acquisitions	FRS	Burke, James	202-452-2612
Competitive Analysis of Mergers and Acquisitions	FRS	Rhoades, Stephen	202-452-3906
Condition Reports	FRS	Carpenter, Douglas	202-452-2740
Consumer Credit, Analysis	FRS	August, James	202-452-3741
Consumer Credit, Analysis	FRS	Luckett, Charles	202-452-2925
Consumer Credit, Data	FRS	Middleton, Millie	202-452-2924
Consumer Finances, Survey	FRS	Fries, Gerhard	202-452-2578
Consumer Leasing	FRS	Hurt, Cecilia	202-452-2412
Credit Cards, Analysis	FRS	Luckett, Charles	202-452-2925
Credit Cards, Data	FRS	Middleton, Millie	202-452-2924
Debt, Domestic Non-Financial	FRS	Holden, Sarah	202-452-3483
Debt, Domestic Non-Financial	FRS	McIntosh, Susan	202-452-3484
Discount Rates at Federal Reserve Banks	FRS	Coyne, Joseph	202-452-3204
Discounts and Advances, Federal Reserve Banks	FRS	Gillum, Gary	202-452-3253
Earnings, Federal Reserve Banks	FRS	Bettge, Paul	202-452-3174
Electric Power Statistics	FRS	Moyers, Cora F.	202-452-2476
Electronic Funds Transfers	FRS	Bowman, Sharon	202-452-3867
Employment Figures	FRS	Lebow, David	202-452-3057
Employment Figures	FRS	Wascher, William	202-452-2812
Energy	FRS	French, Mark	202-452-2348
Energy, International	FRS	Melick, William	202-452-2296
Eurobond Market	FRS	Stekler, Lois	202-452-3716
Eurocurrency Market	FRS	Frankel, Allen	202-452-3578
Excess Reserves	FRS	Ribble, Leigh	202-452-2385
Exchange Rates	FRS	Smith, Ralph	202-452-3712
Exchange Rates, Data	FRS	Decker, Patrick	202-452-3314
Export-Import Bank	FRS	Connors, Thomas	202-452-3639
Export-Trading Companies	FRS	Ryback, William	202-452-2722
Federal Budget Receipts and Outlays	FRS	Follette, Glen	202-452-2448
Federal Budget Receipts and Outlays	FRS	Mariger, Randall	202-452-3703
Federal Budget Receipts and Outlays	FRS	Ramm, Wolfhard	202-452-2381

Part Two: The DATAPHONER   Banking and Financial Data

Subject	Source	Data Contact	Telephone
Federal Reserve Bulletin Tables	FRS	White-Dubose, G.	202-452-3567
Federal Reserve Bulletin Articles	FRS	Dykes, Ellen	202-452-3952
Financial Assets and Liabilities in U.S.	FRS	Fogler, Elizabeth	202-452-3491
Financial Assets and Liabilities in U.S.	FRS	Teplin, Albert M.	202-452-3482
Financial Markets, International	FRS	Smith, Ralph	202-452-3712
Financial Reports	FRS	McLaughlin, Mary	202-452-3829
Foreign Bank Holding Cos. Operating in U.S	FRS	Martinson, M.	202-452-3640
Foreign Banking	FRS	Frankel, Allen	202-452-3578
Foreign Banking Holding Companies in U.S.	FRS	Ryback, William	202-452-2722
Foreign Banks in U.S.	FRS	Ryback, William	202-452-2722
Foreign Banks, In U.S., Economic Analysis	FRS	Key, Sidney	202-452-3522
Foreign Banks with U.S. Branches	FRS	Ryback, William	202-452-2722
Foreign Branches of U.S. Banks	FRS	Roberts, Elizabeth	202-452-3846
Foreign Economies, Africa, Asia, Latin America	FRS	Conners, Thomas	202-452-3639
Foreign Economies, Canada, Europe, Japan	FRS	Johnson, Karen	202-452-2345
Foreign Economies, Europe, Former USSR	FRS	Chang, Valerie	202-452-2375
Foreign Exchange Markets	FRS	Smith, Ralph	202-452-3712
Foreign Exchange Rates	FRS	Decker, Patrick	202-452-3314
Foreign Official Reserves	FRS	Smith, Ralph	202-452-3712
Foreign Trade, U.S.	FRS	Morisse, Kathryn	202-452-3773
Freedom of Information Office	FRS	Mitchell, Susanne	202-452-3684
Futures Trading	FRS	Homer, Laura	202-452-2781
Futures Trading	FRS	Plotkin, Robert	202-452-2782
Gold	FRS	Adams, Donald	202-452-2364
Government Finance	FRS	Simon, Timothy	202-452-2383
Household Balance Sheets and Net Worth	FRS	Fogler, Elizabeth	202-452-3491
Housing Starts and Residential Construction Data	FRS	Whetzel, Frederick	202-452-3094
Housing Starts Residential Construction, Analysis	FRS	Fergus, James	202-452-2868
Industrial Output	FRS	Corrado, Carol	202-452-3521
Industrial Production Index	FRS	Storch, Gerald	202-452-2932
Information System FR Bank Operations	FRS	May, Rick	202-452-3995
Interest Bearing Notes	FRS	Cohen, Sheryl	202-452-3471
Interest on Deposits	FRS	Jorgenson, Harry	202-452-3778
Interest on Federal Reserve Notes	FRS	Bettge, Paul	202-452-3174
Interest on Savings Accounts	FRS	Reid, Brian	202-452-3589
Interest Rates	FRS	Garrett, Bonnie	202-452-2869
Interest Rates, Data	FRS	McMillian, D.	202-452-2851
Interest Rates, Foreign	FRS	Decker, Patrick	202-452-3314
Interest Rates, General Information	FRS	Recording	202-452-6459
International Banking	FRS	Frankel, Allen	202-452-3578

Subject	Source	Data Contact	Telephone
International Banking Facilities	FRS	Key, Sidney	202-452-3522
International Banking Operations	FRS	Ryback, William	202-452-2722
International Capital Transactions	FRS	Stekler, Lois	202-452-3716
International Development	FRS	Connors, Thomas	202-452-3639
International Financial Operations	FRS	Smith, Ralph	202-452-3712
International Information Center	FRS	Sutton, Cynthia	202-452-3411
International Monetary Fund (IMF)	FRS	Connors, Thomas	202-452-3639
International Taxation	FRS	Frankel, Allen	202-452-3578
International Trade, Foreign	FRS	Freeman, Richard	202-452-2344
International Trade Policy, U.S.	FRS	Mann, Catherine	202-452-2374
International Trade Policy, U.S.	FRS	Carper, Virginia	202-452-3661
International Trade, U.S.	FRS	Morisse, Kathryn	202-452-3773
International Transactions, U.S.	FRS	Helkie, William	202-452-3836
Interstate Banking	FRS	Savage, Donald	202-452-2613
Labor Markets and Wages	FRS	Wascher, William	202-452-2812
Labor Markets and Wages	FRS	Otoo, Maria Ward	202-452-2236
Labor Markets and Wages	FRS	Lebow, David	202-452-3057
Manufacturing Capacity	FRS	Gilbert, Charles	202-452-3197
Member Bank Reserves	FRS	Ireland, Oliver	202-452-3625
Member Banks, Trust Activities	FRS	Vinnedge, Donald	202-452-2717
Mergers (Banks and Holding Banks)	FRS	Sussan, Sidney	202-452-2638
Minority Banks	FRS	Monteiro, Anna	202-452-2948
Minority Banks	FRS	Mc Daniel, J.	202-452-3132
Monetary Policy Questions	FRS	Kohn, Donald	202-452-3761
Monetary Policy Stabilization, Special Studies	FRS	Tinsley, P. A.	202-452-2438
Money Supply Deposits and Reserves, U.S.	FRS	Ribble, Leigh	202-452-2385
Mortgage Market Analysis	FRS	Lumpkin, Stephen	202-452-2378
Mortgage Market Data	FRS	Whetzel, Frederick	202-452-3094
Multi-Country Model (MCM)	FRS	Tryon, Ralph	202-452-2368
Municipal Securities Dealer Banks	FRS	Schoenfeld, M.	202-452-2781
National Debt	FRS	Ramm, Wolfhard	202-452-2381
National Debt	FRS	Foliette, Glen	202-452-2448
National Income	FRS	Struckmeyer, S.	202-452-3090
National Information Center	FRS	Hannan, Maureen	202-452-3618
National Information Center	FRS	McDaniel, J.	202-452-3132
Negotiable Orders of Withdrawal (NOW)	FRS	Ribble, Leigh	202-452-2385
Nonbank Acquisitions	FRS	Sussan, Sidney	202-452-2638
NOW Accounts	FRS	Ribble, Leigh	202-452-2385
OECD	FRS	Johnson, Karen	202-452-2345
OECD	FRS	Promisel, Larry	202-452-3533
Open Market Operations	FRS	Garrett, Bonnie	202-452-2869

Subject	Source	Data Contact	Telephone
Open Market Operations	FRS	Simon, Timothy	202-452-2383
OTC (Over-The-Counter Margin Stock List)	FRS	Wolffrum, M.	202-452-2781
Price and Inflation Developments	FRS	Struckmeyer, S.	202-452-3090
Price and Inflation Developments	FRS	Braun, Steven	202-452-3800
Private Pension Funds	FRS	Lander, Joel	202-452-2227
Quarterly Econometric Model	FRS	Brayton, Flint,	202-452-2670
Reports of Condition and Income	FRS	Carpenter, Douglas	202-452-2740
Research Library (FRS)	FRS	Vincent, Susan	202-452-3398
Reserves, Aggregates	FRS	Ribble, Leigh	202-452-2385
Reserves, Computation of	FRS	Bethea, Martha	202-452-3181
Reserves, Foreign	FRS	Smith, Ralph	202-452-3712
Reserves of Depository Institutions	FRS	Ireland, Oliver	202-452-3625
Reserves, Requirements	FRS	Jorgenson, Harry	202-452-3778
Resolution Trust Corporation Oversight Board	FRS	Struble, Frederick	202-452-3794
Savings and Loan Associations	FRS	Passmore, Wayne	202-452-6432
Savings Deposits	FRS	Reid, Brian	202-452-3589
Savings Flow in U.S. Economy	FRS	Fogler, Elizabeth	202-452-3491
Savings Statistics	FRS	Fogler, Elizabeth	202-452-3491
Seasonal Adjustments	FRS	Pierce, David	202-452-3895
State and Local Sector Fiscal Data	FRS	Rubin, Laura	202-452-3130
Statistical Services	FRS	Kim, Po Kyung	202-452-3842
Survey of Consumer Finances	FRS	Fries, Gerhard	202-452-2578
Swap Network	FRS	Smith, Ralph	202-452-3712
T. Bills, Notes, Bonds, Technical Information	FRS	Decorleto, Donna	202-452-3954
Thrift Acquisitions	FRS	McMillian, D.	202-452-2638
Thrift Acquisitions	FRS	Wassom, Molly	202-452-2305
Trade and Financial Studies	FRS	Tryon, Ralph	202-452-2368
Transfer of Funds	FRS	Brett, Gayle	202-452-2934
Treasury Securities, Interest Rates and Yields	FRS	Garrett, Bonnie	202-452-2869
Treasury Securities, Interest Rates and Yields Data	FRS	Sussan, Sidney	202-452-2851
Treasury Securities, Interest Rates and Yields Data	FRS	Culbreth, Leonard	202-452-3853
U.S. BHCs Operating in Foreign Countries	FRS	Martinson, M.	202-452-3640
U.S. International Transactions	FRS	Hooper, Peter	202-452-3426
Unemployment Figures	FRS	Wascher, William	202-452-2812
Unemployment Figures	FRS	Otoo, Maria Ward	202-452-2236
Unemployment Figures	FRS	Lebow, David	202-452-3057
Wage Figures	FRS	Otoo, Maria Ward	202-452-2236
Wage Figures	FRS	Lebow, David	202-452-3057
Wage Figures	FRS	Wascher, William	202-452-2812
World Payments and Economic Activity	FRS	Freeman, Richard	202-452-2344

Business and Industry Data

Subject	Source	Data Contact	Telephone
Business Cycle, Composite Indexes	BEA	Robinson, Charles	202-523-0800
Business Cycle Indicators	BEA	Green, George	202-523-0800
Business Cycle Indicators Data	BEA	Staff	202-523-0500
Business Cycle, Statistical Series	BEA	Young, Mary	202-523-0500
Business Establishment List	BLS	Searson, Michael	202-523-6462
Business Owners	Census	McCutcheon, D.	301-763-5517
Business Statistics, Current	BEA	Staff	202-523-6336
Capital Consumption Allowance	BEA	Musgrave, John	202-523-0837
Capital Expenditures	BEA	Cartwright, David	202-523-0791
Capital Measurement	BLS	Harper, Michael	202-523-6010
Capital Stock	BEA	Musgrave, John	202-523-0837
Communications Statistics	Census	Shoemaker, D.	301-763-2662
Computer Price Index	BEA	Won, Gregory	202-523-5421
Construction Censuses and Surveys	Census	Rappaport, Barry	301-763-5435
Construction Data, Guam	Census	Larson, Odell	301-763-8226
Construction Estimates	BEA	Robinson, Brooks	202-523-0592
Construction Statistics	Census	Visnansky, Bill	301-763-7546
Consumer Expenditure Survey	Census	Hoff, Gail	301-763-2063
Corporate Profits	BEA	Petrick, Kenneth	202-523-0888
Corporate Taxes	BEA	Petrick, Kenneth	202-523-0888
County Business Patterns	Census	Decker, Zigmund	301-763-5430
County Personal Income	BEA	Hazen, Linnea	202-254-6642
Current Industrial Reports (CIRs), Nondurables.	Census	Flood, Thomas	301-763-5911
Current Industrial Reports (CIRs), Durables.	Census	Bernhardt, M.	301-763-2518
Depreciation	BEA	Musgrave, John	202-523-0837
Dividends	BEA	Petrick, Kenneth	202-523-0888
Durable Goods Mfg. Product Data (CIRs)	Census	Bernhardt, M.	301-763-2518
Durables Manufacturing Industry Data	Census	Hansen, Kenneth	301-763-7304
Earnings by Industry, Current Pop. Survey (CPS)	BLS	Stinson, John	202-523-1959
Earnings by Industry, Current Pop. Survey (CPS)	BLS	Parks, William	202-523-1959
Economic Growth	BLS	Bowman, Charles	202-272-5383
Economic Projections	BLS	Saunders, Norman	202-272-5248
Electric Energy Consumed by Manufacturers	Census	McNamee, John	301-763-5938
Employer Cost for Employee Compensation	BLS	Rogers, Brenda	202-523-1165
Employer Cost for Employee Compensation	BLS	Braden, Brad	202-523-1165

Subject	Source	Data Contact	Telephone
Employment by Industry (CPS)	BLS	Parks, William	202-523-1959
Employment Cost Index	BLS	Shelly, Wayne	202-523-1165
Enterprise Statistics	Census	Monaco, Johnny	301-763-1758
Environmental Studies	BEA	Rutledge, Gary	202-523-0687
Establishment Survey, Industrial Classification	BLS	Getz, Patricia	202-523-1172
Establishments, County Business Patterns	Census	Decker, Zigmund	301-763-5430
Exports, Manufacturing	Census	Goldhirsch, Bruce	301-763-1503
Exports, Net, U.S.	BEA	Ehemann, C.	202-523-0669
Farm Income	BEA	Smith, George	202-523-0821
Farm Output	BEA	Smith, George	202-523-0821
Farm Product	BEA	Smith, George	202-523-0821
Farm Proprietors' Income, Subnational Data	BEA	Zavrel, James	202-254-6638
Finance Statistics	Census	Marcus, Sidney	301-763-1386
Financial Data, Quarterly Estimates	Census	Zarrett, Paul	301-763-2718
Fisheries, Annual World Catch Data	NMFS	Holliday, Mark	301-713-2328
Fisheries, Commercial Landings (Catch)	NMFS	Holliday, Mark	301-713-2328
Fisheries, Number of Vessels and Fishermen	NMFS	Holliday, Mark	301-713-2328
Fisheries, Processed Products Data	NMFS	Holliday, Mark	301-713-2328
Fisheries, Shrimp Imports by Country	NMFS	Holliday, Mark	301-713-2328
Food Retailing, Agriculture	ERS	Handy, Charles	202-219-0866
Food Wholesaling, Agriculture	ERS	Epps, Walter	202-219-0866
Foreign Direct Investment, Benchmark Data	BEA	Bezirganian, Steve	202-523-0641
Foreign Direct Investment, Annual Survey	BEA	Bezirganian, Steve	202-523-0641
Fuels Consumed by Manufacturers	Census	McNamee, John	301-763-5938
Gross Domestic Product, Current Estimates	BEA	Mannering, V.	202-523-0824
Gross Domestic Product by Industry	BEA	Mohr, Michael	202-523-0795
Gross National Product, Current Estimates	BEA	Mannering, V.	202-523-0824
Gross Private Domestic Investment	BEA	Cartwright, David	202-523-0791
Gross State Product Estimates	BEA	Dunbar, Ann	202-523-9180
Guam, Construction Data	Census	Larson, Odell	301-763-8226
Hourly Earnings Index	BLS	Braden, Brad	202-523-1165
Households, Wealth	Census	Lamas, Enrique	301-763-8578
Industry Classification Information	Census	Venning, Alvin	301-763-1935
Industry Data, Nondurables Manufacturing	Census	Zampogna, M.	301-763-2510
Industry Projections	BLS	Bowman, Charles	202-272-5383
Industry Statistics	Census	Priebe, John	301-763-8574
Industry Statistics	Census	Masumura, Will	301-763-8574
Industry-Occupation Employment Matrix	BLS	Turner, Delores	202-272-5383
Input-Output and Industry Data	BLS	Franklin, James	202-272-5240
Input-Output and Industry Data	BLS	Andreassen, Art	202-272-5326
Input-Output Annual Tables	BEA	Planting, Mark	202-523-0867

Business and Industry Data

XI. Business Data Directory

Subject	Source	Data Contact	Telephone
Input-Output Benchmark Tables	BEA	Maley, Leo	202-523-0683
Input-Output Tables, Goods Industries	BEA	Bonds, Belinda	202-523-0843
Input-Output Tables, Service Industries	BEA	Horowitz, Karen	202-523-3505
Insurance Statistics	Census	Marcus, Sidney	301-763-1386
Interest Income	BEA	Weadock, Teresa	202-523-0833
Interest Payments	BEA	Weadock, Teresa	202-523-0833
Interest, Subnational Estimates	BEA	Jolley, Charles	202-254-6637
Inventories	BEA	Baldwin, Steve	202-523-0784
Inventories, Monthly Retail Trade Data	Census	Piencykoski, Ron	301-763-5294
Inventory Sales Ratios	BEA	Stiller, Jean	202-523-6585
Inventory Statistics, Wholesale Trade	Census	Gordon, Dale	301-763-3916
Investment in Plant and Equipment	Census	Gates, John	301-763-5596
Labor Composition and Hours Worked Survey	BLS	Rosenblum, Larry	202-523-9261
Manufacturing Capacity	Census	Champion, Elinor	301-763-5616
Manufacturing Concentration	Census	Goldhirsch, Bruce	301-763-1503
Manufacturing Exports	Census	Goldhirsch, Bruce	301-763-1503
Manufacturing, Food	ERS	Handy, Charles	202-219-0866
Manufacturing, Guam	Census	Hoover, Kent	301-763-8564
Manufacturing Industry Statistics	Census	Govoni, John	301-763-7666
Manufacturing Inventories, Monthly Data	Census	Runyan, Ruth	301-763-2502
Manufacturing Orders, Monthly Data	Census	Runyan, Ruth	301-763-2502
Manufacturing Pollution Abatement	Census	Shapiro, Janet	301-763-1755
Manufacturing Production Index	Census	Goldhirsch, Bruce	301-763-1503
Manufacturing, Puerto Rico	Census	Larson, Odell	301-763-8226
Manufacturing Research and Development	Census	Champion, Elinor	301-763-5616
Manufacturing Shipments, Monthly Data	Census	Runyan, Ruth	301-763-2502
Manufacturing, Virgin Islands	Census	Hoover, Kent	301-763-8564
Marketing Margins and Statistics, Agriculture	ERS	Haidacher, R.	202-219-0870
Metropolitan Area Personal Income	BEA	Brown, Robert	202-254-6632
Mineral Industry Statistics	Census	McNamee, John	301-763-5938
Minority-Owned Businesses	Census	McCutcheon, D.	301-763-5517
Monthly Manufacturing Inventories	Census	Runyan, Ruth	301-763-2502
Monthly Manufacturing Orders	Census	Runyan, Ruth	301-763-2502
Monthly Manufacturing Shipments	Census	Runyan, Ruth	301-763-2502
Motor Freight Transportation	Census	Zabelsky, Tom	301-763-1725
Multifactor Productivity	BLS	Rosenblum, Larry	202-523-9261
National Income	BEA	Seskin, Eugene	202-523-0848
New Investment Survey, Analysis	BEA	Fahim-Nader, M.	202-523-0640
New Investment Survey Data	BEA	Cherry, Joseph	202-523-0654
Nondurable Mfg., Census, Annual Survey	Census	Zampogna, Mike	301-763-2510
Nondurables Mfg., Current Industrial Reports	Census	Flood, Thomas	301-763-5911

Subject	Source	Data Contact	Telephone
Nonfarm Proprietors' Income, Subnational	BEA	Levine, Bruce	202-254-6634
OSHA Recordkeeping Requirements	BLS	Whitmore, Robert	202-272-3462
Output Measures	BEA	Ehemann, Chris	202-523-0669
Personal Consumption Expenditures, Services	BEA	McCully, Clint	202-523-0819
Personal Consumption Expenditures, Prices	BEA	McCully, Clint	202-523-0819
Personal Consumption Expenditures, Autos	BEA	Johnson, Everette	202-523-0807
Personal Consumption Expenditures, Other	BEA	Key, Greg	202-523-0778
Personal Income in Metropolitan Areas	BEA	Brown, Robert	202-254-6632
Plant and Equipment Investment	Census	Gates, John	301-763-5596
Plant and Equipment Expenditures	BEA	Crawford, Jeffrey	202-523-0782
Pollution Abatement Activities in Manufacturing	Census	Shapiro, Janet	301-763-1755
Pollution Abatement and Control Spending	BEA	Rutledge, Gary	202-523-0687
Price Measures, Chain Price Indexes	BEA	Herman, Shelby	202-523-0828
Price Measures, Fixed-Weighted Indexes	BEA	Herman, Shelby	202-523-0828
Prices of Personal Consumption Expenditures	BEA	McCully, Clint	202-523-0819
Producers Durable Equipment Data	BEA	Crawford, Jeffrey	202-523-0782
Product Data, Nondurables Manufacturing	Census	Zampogna, Mike	301-763-2510
Productivity, Multifactor	BLS	Rosenblum, Larry	202-523-9261
Productivity Trends in Selected Industries	BLS	Ardolini, Charles	202-523-9244
Projections, Economic, States / Metropolitan Areas	BEA	Johnson, Kenneth	202-523-0971
Proprietors' Income, Nonfarm	BEA	Abney, Willie	202-523-0811
Puerto Rico, Construction Data	Census	Larson, Odell	301-763-8226
Quarterly Financial Report	Census	Zarrett, Paul	301-763-2718
Real Estate Statistics	Census	Marcus, Sidney	301-763-1386
Regional Economic Situation, Current	BEA	Friedenberg, H.	202-523-0979
Regional Input-Output Modeling System (RIMS)	BEA	Pigler, Carmen	202-523-0586
Rental Income	BEA	Smith, George	202-523-0821
Rental Income, Subnational Estimates	BEA	Jolley, Charles	202-254-6637
Research and Development in Manufacturing	Census	Champion, Elinor	301-763-5616
Retail Trade, Advance Monthly Sales	Census	Piencykoski, Ron	301-763-5294
Retail Trade, Annual Data	Census	Piencykoski, Ron	301-763-5294
Retail Trade, Census	Census	Russel, Anne	301-763-7038
Retail Trade, Guam	Census	Hoover, Kent	301-763-8564
Retail Trade, Monthly Report	Census	True, Irving	301-763-7128
Retail Trade, Monthly Inventories	Census	Piencykoski, Ron	301-763-5294
Retail Trade, Puerto Rico	Census	Larson, Odell	301-763-8226
Retail Trade, Virgin Islands	Census	Hoover, Kent	301-763-8564
Sales, Advance Monthly Retail Trade Data	Census	Piencykoski, Ron	301-763-5294
Sales, Annual Retail Trade Data	Census	Piencykoski, Ron	301-763-5294
Sales Statistics, Wholesale Trade	Census	Gordon, Dale	301-763-3916
Saving	BEA	Donahoe, Gerald	202-219-0669

Business and Industry Data   *XI. Business Data Directory*

Subject	Source	Data Contact	Telephone
Service Industries Census	Census	Moody, Jack	301-763-7039
Service Industry, Current Statistics	Census	Zabelsky, Tom	301-763-1725
Service Industry Data and Reports	Census	Zabelsky, Tom	301-763-1725
Services, Guam	Census	Larson, Odell	301-763-8226
Services, Puerto Rico	Census	Hoover, Kent	301-763-8564
Services, Virgin Islands	Census	Larson, Odell	301-763-8226
Shift-Share Analysis	BEA	Kort, John	202-523-0946
Shippers' Export Information	Census	Blyweiss, Hal	301-763-5310
State Econometric Modeling	BEA	Lienesch, Thomas	202-523-0943
State Personal Income, Quarterly Data	BEA	Whiston, Isabelle	202-254-6672
State Personal Income, Annual Data	BEA	Hazen, Linnea	202-254-6642
Structures	BEA	Robinson, Brooks	202-523-0592
Taxes, Corporate	BEA	Petrick, Kenneth	202-523-0888
Technology Trends in Major Industries	BLS	Riche, Richard	202-523-9311
Trade, Merchandise, International Data	BEA	Murad, Howard	202-523-0668
Transportation Statistics	Census	Shoemaker, D.	301-763-2662
Travel Surveys	Census	Cannon, John	301-763-5468
Truck Inventory and Use	Census	Bostic, William	301-763-2735
Trucking Services	Census	Zabelsky, Tom	301-763-1725
U.S. Direct Investment Abroad, Analysis	BEA	Mataloni, Ray	202-523-3451
Utility Statistics	Census	Shoemaker, D.	301-763-2662
Value of New Construction Put in Place	Census	Meyer, Allan	301-763-5717
Virgin Islands, Construction Data	Census	Larson, Odell	301-763-8226
Wages by Industry, State and Area	BLS	Bush, Joseph	202-523-1158
Warehousing Services	Census	Zabelsky, Tom	301-763-1725
Water Use by Manufacturers	Census	McNamee, John	301-763-5938
Wealth Estimates	BEA	Musgrave, John	202-523-0837
Wholesale Trade Census	Census	Trimble, John	301-763-5281
Wholesale Trade, Current Sales Statistics	Census	Gordon, Dale	301-763-3916
Wholesale Trade, Current Inventory Statistics	Census	Gordon, Dale	301-763-3916
Wholesale Trade, Guam	Census	Hoover, Kent	301-763-8564
Wholesale Trade, Puerto Rico	Census	Larson, Odell	301-763-8226
Wholesale Trade, Virgin Islands	Census	Hoover, Kent	301-763-8564
Wholesaling, Food	ERS	Epps, Walter	202-219-0866
Women-Owned Businesses	Census	McCutcheon, D.	301-763-5517
Zip Code Economic Data	Census	Russell, Anne	301-763-7038

Data Managers

Subject	Source	Data Contact	Telephone
Agricultural Data	Census	Pautler, P. Charles	301-763-8555
Agricultural Data Systems and Information	NASS	Zellers, Phillip	202-720-2984
Agricultural Estimates	NASS	Vogel, Frederic	202-720-3896
Agricultural Research	NASS	Clark, Cynthia	202-720-4557
Balance of Payments	BEA	Bach, Christopher	202-523-0620
Business Data, Trade, Services, Finance	Census	Hamilton, Howard	301-763-7564
Business Establishment Systems	BLS	Carlson, Robert	202-501-6795
Business Outlook	BEA	Green, George	202-523-0701
Census Information Services	Census	DiCesare, C.	202-523-1090
Center for Economic Studies	Census	McGuckin, Bob	301-763-2337
Center for International Research	Census	Torrey, Barbara B.	301-763-2870
Center for Survey Methods Research	Census	Martin, Elizabeth	301-763-3838
Compensation and Working Conditions	BLS	Stelluto, George	202-272-1382
Compensation and Working Conditions Programs	BLS	McNulty, Donald	202-523-1228
Computer Systems and Services	BEA	Doyle, James	202-523-0978
Construction Statistics	Census	Richardson, Joel	301-763-7163
Consumer Expenditure Surveys	BLS	Jacobs, Eva	202-272-5156
Consumer Expenditures, Statistical Methods	BLS	Hsen, Paul	202-272-2321
Consumer Price Index Data Collection Research	BLS	Uglow, David	202-272-2323
Consumer Price Index Quality	BLS	Branscome, Jim	202-272-2322
Consumer Price Index Survey Research	BLS	Williams, Janet	202-272-2281
Consumer Prices	BLS	Lane, Walter	202-272-3583
Consumer Prices and Consumption Studies	BLS	Wright, Stephen	202-272-5002
Consumer Prices and Price Indexes	BLS	Armknecht, Paul	202-272-5164
Current Business Analysis	BEA	Fox, Douglas	202-523-0697
Current Employment Analysis	BLS	Bregger, John	202-523-1824
Data User Services	Census	Turner, Marshall	301-763-5820
Decennial Operations	Census	Jackson, Arnold	301-763-2682
Demographic Surveys	Census	Courtland, Sherry	301-763-2776
Directly Collected Periodic Surveys	BLS	Richards, Gregory	202-272-5483
Durable Goods Industry Prices	BLS	Lavish, Kenneth	202-272-5115
Economic Census Staff	Census	Messenbourg, Tom	301-763-7356
Economic Programming	Census	Cohen, Barry	301-763-2912
Economic Surveys	Census	Staff	301-763-7735
Employment and Unemployment Data	BLS	Ziegler, Martin	202-523-1919
Employment / Unemployment Statistical Methods	BLS	Tupek, Alan	202-523-1695

Data Managers — XI. Business Data Directory

Subject	Source	Data Contact	Telephone
Employment Cost Trends	BLS	Wood Jr, Donald	202-523-1160
Employment Statistics by Month and Industry	BLS	Werking, George	202-523-1446
Environmental Economics	BEA	Rutledge, Gary	202-523-0687
Federal and State Monthly Surveys	BLS	Powers, Brendan	202-523-1001
Foreign Labor Statistics	BLS	Neef, Arthur	202-523-9291
Foreign Trade Data	Census	Adams, Don	301-763-5342
Foreign Trade Data Systems	Census	Woods, Charles	301-763-7982
Geography	Census	Marx, Robert	301-763-5636
Government Data	Census	Green, Gordon	301-763-7366
Government Data	BEA	Ziemer, Richard	202-523-0715
Housing and Household Economic Statistics	Census	Weinberg, Daniel	301-763-8550
Industrial Price Information and Current Analysis	BLS	Howell, Craig	202-272-5113
Industrial Prices and Price Indexes	BLS	Tibbetts, Thomas	202-272-5110
Industrial Prices and Relations	BLS	Kirsch, Philip	202-272-5182
Industry Data	Census	Worden, Gaylord	301-763-5850
Industry Employment Projections	BLS	Bowman, Charles	202-272-5383
Industry Productivity Studies	BLS	Ardolini, Charles	202-523-9244
Information Services	BLS	DiCesare, C.	202-523-1090
Interindustry Economics	BEA	Maley, Leo	202-523-0683
International Investment Data	BEA	Barker, Betty	202-523-0659
International Price Index Methods	BLS	Alterman, William	202-272-5020
International Price Indexes	BLS	McCarthy, Mary	202-272-5026
International Price Systems	BLS	Share, David	202-504-2158
International Prices	BLS	Kasper, Marvin	202-272-2272
International Statistical Programs Center	Census	Walsh, Thomas	301-763-2832
International Technical Assistance	BLS	McCracken, John	202-523-9231
Labor Force Programs, Testing and Evaluation	BLS	Hines, Joseph	202-504-2020
Labor Force Statistics	BLS	Flaim, Paul	202-523-1944
Labor-Management Relations, Developments	BLS	Bauman, Alvin	202-523-1143
Local Area Unemployment Statistics	BLS	Brown, Sharon	202-523-1038
National Income and Wealth Data	BEA	Donahoe, Gerald	202-523-0669
Non-Durable Goods Industry Prices	BLS	Gaddie, Robert	202-272-5137
Occupational and Administrative Statistics	BLS	MacDonald, Brian	202-523-1949
Occupational Outlook	BLS	Rosenthal, Neal	202-272-5382
Occupational Pay and Employee Benefit Levels	BLS	Pfuntner, Jordon	202-523-1246
Population Data	Census	Schneider, Paula	301-763-7646
Price and Index Number Research	BLS	Zieschang, Kim	202-272-5096
Price Index Analysis	BLS	Rosenberg, Elliott	202-272-5118
Price Index Information and Current Analysis	BLS	Howell, Craig	202-272-5113
Price Index Methods, Analysis and Evaluation	BLS	Tschetter, M.	202-272-5170
Price Programs	BLS	Jack, Patricia	202-504-2015

Subject	Source	Data Contact	Telephone
Producer Price Index	BLS	Pratt, Richard	202-272-2196
Productivity Research	BLS	Harper, Michael	202-523-6010
Program and Policy Development Office	Census	Miller, Catherine	301-763-2758
Public Information	Census	Cagle, Maury	301-763-4040
Regional Economic Analysis	BEA	Kort, John	202-523-0946
Regional Economic Data	BEA	Hazen, Linnea	202-254-6642
Safety and Health Systems	BLS	Chen-Nash, Elaine	202-501-6448
Safety, Health, Program Analysis	BLS	Weber, William	202-501-6468
Service Industry Prices	BLS	Gerduk, Irwin	202-272-5130
State Agricultural Statistics	NASS	Barrett, Fred	202-720-3638
Statistical Methods	BLS	Hedges, Brian	202-272-2195
Statistical Methods	Census	Waite, Preston	301-763-2672
Statistical Research	Census	Tortora, Robert	301-763-3807
Statistical Support	Census	Thompson, John	301-763-4072
Wage Statistical Methods	BLS	Cohen, Stephen	202-523-5922
Year 2000 Research and Development	Census	Miskura, Susan	301-763-8601

Data Products and Other Information Services

Subject	Source	Data Contact	Telephone
Agricultural Data, Current	NASS	Staff	202-720-4020
Agricultural Reports and Publications	NASS	Staff	202-720-4020
Agriculture, General Information	Census	Towers, Sharon	800-523-3215
BEA Economic Data on Electronic Bulletin Board	OBA	Staff	202-377-1986
BLS Data Diskettes and Tapes	BLS	Buso, Michael	202-523-1158
BLS Indicators	BLS	Recorded Message	202-523-9658
BLS Press Officer	BLS	Hoyle, Kathryn	202-523-1913
BLS Publications Information	BLS	Recorded Message	202-523-1221
Business and Industry Data Centers Information	Census	Rowe, John	301-763-1580
CD-ROM Information	Census	Staff	301-763-4673
CENDATA Services	Census	Staff	301-763-2074
Census Count Information, Decennial	Census	Staff	301-763-5002
Census Publications, Decennial	Census	Landman, Cheryl	301-763-3938
Census Tabulations, Decennial	Census	Porter, Gloria	301-763-4908
Clearinghouse for Census Data Services	Census	Staff	301-763-1580
Computerized Agricultural Data	NASS	Staff	202-720-6306
Computerized Economic Data	BEA	Howenstine, B.	202-523-0777
Computerized Economic Data	Census	Staff	301-763-4100
Computerized Labor Data	BLS	Ayres, Mary Ellen	202-523-7827
Consumer Price Index	BLS	Recorded Message	202-523-1221
Consumer Price Index Data Diskettes	BLS	Gibson, Sharon	202-504-2051
Consumer Price Index Details	BLS	Recorded Message	202-523-1221
County and City Data Books	Census	Cevis, Wanda	301-763-1034
Digital Map Data Base (TIGER)	Census	Staff	301-763-1580
Earnings, State and Local, Data Diskettes	BLS	Podgornik, Guy	202-523-1759
Economic Census Products	Census	Zeisset, Paul	301-763-1792
Economic Data, Current - In Los Angeles, CA	Census	Staff	818-904-6339
Economic Data, Current - In Denver, CO	Census	Staff	303-969-7750
Economic Data, Current - In Washington, DC	BEA	Howenstine, B.	202-523-0777
Economic Data, Current - In Washington, DC	Census	Staff	301-763-4100
Economic Data, Current - In Atlanta, GA	Census	Staff	404-347-2274
Economic Data, Current - In Chicago, IL	Census	Staff	312-353-0980
Economic Data, Current - In Kansas City, KS	Census	Staff	816-891-7562
Economic Data, Current - In Boston, MA	Census	Staff	617-565-7078
Economic Data, Current - In Detroit, MI	Census	Staff	313-354-4654
Economic Data, Current - In New York, NY	Census	Staff	212-264-4730

Subject	Source	Data Contact	Telephone
Economic Data, Current - In Charlotte, NC	Census	Staff	704-371-6142
Economic Data, Current - In Philadelphia, PA	Census	Staff	215-597-8313
Economic Data, Current - In Dallas, TX	Census	Staff	214-767-7105
Economic Data, Current - In Seattle, WA	Census	Staff	206-728-5314
Economic Data on Electronic Bulletin Board	Census	Staff	301-763-1580
Economic Data on Electronic Bulletin Board	OBA	Staff	202-377-1986
Economic Reports and Publications	BEA	Howenstine, B.	202-523-0777
Economic Reports and Publications	Census	Staff	301-763-4100
Electronic Bulletin Board, Economic Data	Census	Staff	301-763-1580
Electronic Bulletin Board, Economic Data	OBA	Staff	202-377-1986
Employment and Earnings Publication	BLS	Green, Gloria	202-523-1959
Employment Cost Index Diskettes	BLS	Rogers, Brenda	202-523-1165
Employment Cost Index Information	BLS	Recorded Message	202-523-1221
Employment Data, State and Local, Data Diskettes	BLS	Podgornik, Guy	202-523-1759
Employment Projections Tapes and Diskettes	BLS	Bowman, Charles	202-272-5383
Employment Situation, News Release	BLS	Staff	202-523-1944
Employment Situation, Recorded Message	BLS	24-Hour Hot Line	202-523-1221
Employment, Subnational Data	BEA	Staff	202-254-6630
Establishment Survey Data Diskettes	BLS	Hiles, David	202-523-1172
Foreign Trade Data Inquiries	Census	Higbee, Reba	301-763-5140
Foreign Trade Data Services	Census	Mearkle, Haydn	301-763-7754
Foreign Trade Press Releases and Publications	Census	Preuss, Richard	301-763-7754
Gross National Product	BEA	Recorded Message	202-898-2451
Gross National Product Computer Tapes	BEA	Blue, Eunice	202-523-0804
Gross National Product Diskettes	BEA	Blue, Eunice	202-523-0804
Gross National Product Printouts	BEA	Blue, Eunice	202-523-0804
Historical Census Data	Census	Staff	301-763-7936
Hours Worked, State and Local, Data Diskettes	BLS	Podgornik, Guy	202-523-1759
Income and Program Participation, Survey (SIPP)	Census	Campbell, Carmen	301-763-2005
Input-Output and Industry Data Tapes	BLS	Andreassen, Art	202-272-5326
Input-Output Tables, Computer Tapes	BEA	Carter, Esther	202-523-0792
Input-Output Tables, Diskettes	BEA	Carter, Esther	202-523-0792
Input-Output Tables, Printouts	BEA	Carter, Esther	202-523-0792
Labor Data, Current	BLS	Recorded Message	202-523-1221
Labor Data, Current - In Los Angeles, CA	BLS	Staff	213-252-7521
Labor Data, Current - In San Francisco, CA	BLS	Staff	415-774-6600
Labor Data, Current - In Washington, DC	BLS	Staff	202-523-1221
Labor Data, Current - In Atlanta, GA	BLS	Staff	404-347-4416
Labor Data, Current - In Chicago, IL	BLS	Staff	312-353-1880
Labor Data, Current - In Boston, MA	BLS	Staff	617-565-2327
Labor Data, Current - In New York City, NY	BLS	Staff	212-337-2400

Subject	Source	Data Contact	Telephone
Labor Data, Current - In Philadelphia, PA	BLS	Staff	215-596-1154
Labor Data, Current - In Dallas, TX	BLS	Staff	214-767-6970
Labor Force Machine-Readable Data and Disks	BLS	Green, Gloria	202-523-1959
Labor Force Micro-Data Tapes	BLS	McIntire, Robert	202-523-1776
Labor Related Information	BLS	Wallace, Charles	202-523-1208
Labor Reports and Publications	BLS	Staff	202-523-1221
Leading Indicators	BEA	Recorded Message	202-898-2450
Map Orders, 1980 Census	Census	Baxter, Leila	812-288-3192
Map Orders, 1990 Census	Census	Staff	301-763-4100
Merchandise Trade	BEA	Recorded Message	202-898-2453
National Services Information Centers	Census	Johnson, Sam	301-763-1384
Occupational Injury, Annual Survey, Data Disks	BLS	Jackson, Ethel	202-501-6470
Occupational Injury, Annual Survey, Data Tapes	BLS	Jackson, Ethel	202-501-6470
Occupational Outlook Handbook	BLS	Pilot, Michael	202-272-5382
Personal Income and Outlays	BEA	Recorded Message	202-898-2452
Personal Income, Subnational Data	BEA	Staff	202-254-6630
Producer Price Index	BLS	Recorded Message	202-523-1221
Producer Price Index Data Diskettes	BLS	Rosenberg, Elliott	202-272-5118
Productivity and Costs News Releases	BLS	Fulco, Lawrence	202-523-9261
Productivity Data Diskettes	BLS	Fulco, Lawrence	202-523-9261
Productivity Data Tapes	BLS	Kriebel, Bertram	202-523-9261
Public-Use Microdata Samples	Census	Campbell, Carmen	301-763-2005
Real Earnings, News Release	BLS	Hiles, David	202-523-1172
State Agricultural Data	NASS	State Statistician Offices*	
State and Area Labor Force Diskettes and Tapes	BLS	Marcus, Jessie	202-523-1002
State and Metropolitan Area Data Books	Census	Cevis, Wanda	301-763-1034
State Data Center Program	Census	Carbaugh, Larry	301-763-1580
Statistical Abstract	Census	King, Glenn	301-763-5299
Statistical Briefs	Census	Bernstein, Robert	301-763-1584
Telecommunication Device for the Deaf	BLS	----------------------	202-523-3926

* See National Agricultural Statistics Service State Statisticians' Offices in Section III, State and Regional Data Centers, of the Business Data Directory, for state contacts and telephone numbers.

Employment and Labor Force Data

Subject	Source	Data Contact	Telephone
Commuting, Place of Work	Census	Boertlein, Celia	301-763-3850
County Employment	BEA	Hazen, Linnea	202-254-6642
Demographic Characteristics, State and Area	BLS	Biederman, Edna	202-523-1002
Earnings, Based on Establishment Survey	BLS	Seifert, Mary Lee	202-523-1172
Educational Attainment	BLS	Meisenheimer, Joe	202-523-1944
Employment	Census	Palumbo, Thomas	301-763-8574
Employment	Census	Jones, Selwyn	301-763-8574
Employment, Based on Establishment Survey	BLS	Seifert, Mary Lee	202-523-1172
Employment by Industry, State and Area	BLS	Bush, Joseph	202-523-1158
Employment Cost Index	BLS	Shelly, Wayne	202-523-1165
Employment, Counties	BEA	Hazen, Linnea	202-254-6642
Employment, County Business Patterns	Census	Decker, Zigmund	301-763-5430
Employment Data by State and Area	BLS	Shipp, Kenneth	202-523-1227
Employment Estimates, State	BEA	Hazen, Linnea	202-254-6642
Employment, Metropolitan Areas	BEA	Hazen, Linnea	202-254-6642
Employment, Quarterly Data (ES-202 Report)	BLS	Bush, Joseph	202-523-1158
Employment, State and Local Area	BEA	Levine, Bruce	202-254-6634
Employment, State Estimates	BEA	Carnevale, Sharon	202-254-7703
Establishment Survey Benchmark	BLS	Cronkhite, Fred	202-523-1146
Establishment Survey, State and Area Data	BLS	Shipp, Kenneth	202-523-1227
Family Characteristics of Workers	BLS	Hayghe, Howard	202-523-1371
Family Labor Force Data	BLS	Haughe, Howard	202-523-1371
Farm Employment, Subnational Data	BEA	Zavrel, James	202-254-6638
Farm Labor	ERS	Oliveira, Victor	202-219-0033
Farm Labor	NASS	Kurtz, Tom	202-690-3228
Farm Labor	ERS	Duffield, James	202-219-0033
Farm Labor Laws	ERS	Runyon, Jack	202-219-0932
Farm Labor Market	ERS	Duffield, James	202-219-0033
Farm Labor Market	ERS	Oliveira, Victor	202-219-0033
Government Employment	Census	Stevens, Alan	301-763-5086
Hours Worked, Based on Establishment Survey	BLS	Seifert, Mary Lee	202-523-1172
Industry Employment, CPS Monthly Survey	BLS	Parks, William	202-523-1959
Industry Generated Employment	BLS	Andreassen, Art	202-272-5326
Industry Statistics	Census	Priebe, John	301-763-8574
Job Tenure	BLS	Hamel, Harvey	202-523-1371
Job Vacancies	BLS	Deven, Richard	202-523-1694

Subject	Source	Data Contact	Telephone
Journey to Work Data	Census	Salopek, Phil	301-763-3850
Labor Force Data, Concepts and Definitions	BLS	Stinson, John	202-523-1959
Labor Force Projections	BLS	Fullerton, Howard	202-272-5328
Longitudinal Data, Gross Flows	BLS	Horvath, Francis	202-523-1371
Marital Characteristics of Workers	BLS	Hayghe, Howard	202-523-1371
Metropolitan Area Employment	BEA	Hazen, Linnea	202-254-6642
Minority Employment Data	BLS	Hamel, Harvey	202-523-1371
Minority Labor Force Data	BLS	Cattan, Peter	202-523-1944
Nonfarm Proprietors, Subnational Data	BEA	Zavrel, James	202-254-6638
Occupation Statistics	Census	Masumura, Will	301-763-8574
Occupation Statistics	Census	Priebe, John	301-763-8574
Occupational Data, Current Population Survey	BLS	Nardone, Thomas	202-523-1944
Occupational Employment (OESS)	BLS	Cromartie, Stella	202-523-1371
Occupational Employment Statistics Survey-OESS	BLS	Hadlock, Paul	202-523-1242
Occupational Mobility	BLS	Rones, Philip	202-523-1944
Occupational Outlook Quarterly	BLS	Fountain, Melvin	202-272-5298
Occupational Projections	BLS	Rosenthal, Neal	202-272-5382
Unemployment, County Business Patterns	Census	Decker, Zigmund	301-763-5430
Veterans Employment Statistics	BLS	Cohany, Sharon	202-523-1944
Work Experience	BLS	Mellor, Earl	202-523-1371
Workers, Older	BLS	Rones, Philip	202-523-1944
Workers, Part-Time	BLS	Nardone, Thomas	202-523-1944
Working Poor	BLS	Herz, Diane	202-523-1944
Youth, Students, and Dropouts	BLS	Cohany, Sharon	202-523-1944

Federal, State and Local Government Data

Subject	Source	Data Contact	Telephone
Annexations	Census	Goodman, Nancy	301-763-3827
Apportionment	Census	Speaker, Robert	301-763-7962
Budget, Cyclically-Adjusted	BEA	Webb, Michael	202-523-3470
Congressional District Address Locations	Census	Swapshur, Ernie	301-763-5692
Congressional District Boundaries, Components	Census	Hamill, Robert	301-763-5720
Contributions, Federal Government	BEA	Tsehaye, Benyam	202-523-0885
Criminal Justice Statistics	Census	Cull, Diana	301-763-7789
Employment, Government	Census	Stevens, Alan	301-763-5086
Expenditures, Federal Government	BEA	Dobbs, David	202-523-0744
Expenditures, Federal Government	Census	Kellerman, David	301-763-5276
Expenditures, State and Local Government	BEA	Sullivan, David	202-523-0725
Farm Taxes	ERS	Durst, Ron	202-219-0896
Financial Data, Government	Census	Wulf, Henry	301-763-7664
Legal Areas, Annexations, Boundary Changes	Census	Goodman, Nancy	301-763-3827
National Defense Purchases, Goods and Services	BEA	Galbraith, Karl	202-523-3472
Nondefense Purchases of Goods and Services	BEA	Mangan, Robert	202-523-5017
Operations Support and Analysis, Government	Census	Fanning, William	301-763-4403
Organization Data, Government	Census	Cull, Diana	301-763-7789
Productivity, Government	BLS	Forte, Darlene	202-523-9156
Productivity Trends in Federal Government	BLS	Ardolini, Charles	202-523-9244
Purchases of Goods and Services, State / Local Govt.	BEA	Peters, Donald	202-523-0726
Reapportionment	Census	Turner, Marshall	301-763-5820
Receipts, Federal Government	BEA	Dobbs, David	202-523-0744
Receipts, State and Local Government	BEA	Sullivan, David	202-523-0725
Redistricting	Census	Talbert, Cathy	301-763-4070
Tax Data	Census	Keffer, Gerald	301-763-5356
Transactions, Government	BEA	Thomas, Gregory	202-523-0615
Transfers, Federal Government	BEA	Tsehaye, Benyam	202-523-0885
Voting and Registration Data	Census	Jennings, Jerry	301-763-4547
Voting Districts	Census	McCully, Cathy	301-763-3827
Wages, State and Local Governments	BLS	Field, Charles	202-523-1570

Geographic Concepts and Products

Subject	Source	Data Contact	Telephone
Cartographic Operations	Census	Staff	301-763-3973
Computer-Readable Map and Geographic Data Base	Census	Carbaugh, Larry	301-763-1581
Map Information and Products	Census	Staff	301-763-4100
Mapping, Computer	Census	Broome, Fred	301-763-3973
Metropolitan Areas, (PMSAs, MSAs)	Census	Forstall, Richard	301-763-5158
Metropolitan Areas (PMSAs, MSAs)	Census	Fitzsimmons, Jim	301-763-5158
Neighborhood Statistics (User Defined)	Census	Quasney, A.	301-763-4282
Statistical Areas	Census	Staff	301-763-3827
TIGER Systems Products (Digitized Mapping)	Census	Staff	301-763-1581
Voting Districts	Census	McCully, Cathy	301-763-3827
Zip Code, Geographic Relationships	Census	Quarato, Rose	301-763-4667

Housing Data

Subject	Source	Data Contact	Telephone
American Housing Survey	Census	Montfort, Edward	301-763-8551
Completions, Housing	Census	Fondelier, David	301-763-5731
Construction Authorized by Building Permits	Census	Hoyle, Linda	301-763-7244
Construction Censuses and Surveys	Census	Rappaport, Barry	301-763-5435
Decennial Censuses, Housing Data	Census	Downs, Bill	301-763-8553
Expenditures, Residential Improvements / Repairs	Census	Roff, George	301-763-5705
Group Quarters Population	Census	Smith, Denise	301-763-7883
Housing Markets, Finances, Absorption	Census	Smoler, Anne	301-763-8552
Housing Markets, Finances, Absorption	Census	Fronzek, Peter	301-763-8552
Housing Starts	Census	Fondelier, David	301-763-5731
Inventories, Housing	Census	Maynard, Jane	301-763-8551
New York City Housing and Vacancy Survey	Census	Harper, Margaret	301-763-8552
Residential Construction	BEA	Robinson, Brooks	202-523-0592
Residential Construction in Selected MSAs, New	Census	Jacobson, Dale	301-763-7842
Residential Construction, New	Census	Berman, Steve	301-763-7842
Residential Construction Price Index, New	Census	Berman, Steve	301-763-7842
Residential Construction Sales, New	Census	Berman, Steve	301-763-7842
Residential Finances	Census	Smoler, Anne	301-763-8552
Residential Finances	Census	Fronzek, Peter	301-763-8552
Residential Improvements and Repairs	Census	Roff, George	301-763-5705
Special Tabulations of Housing Data	Census	Downs, Bill	301-763-8553
Urban and Rural Residence Data	Census	Staff	301-763-7962
Vacancy Data	Census	Fraser, Wallace	301-763-8165
Value of New Construction Put in Place	Census	Meyer, Allan	301-763-5717

International and Foreign Trade Data

Subject	Source	Data Contact	Telephone
Africa	ERS	Kurtzig, Michael	202-219-0680
Africa	Census	Hobbs, Frank	301-763-4221
Agricultural Export Programs	ERS	Ackerman, Karen	202-219-0820
Agricultural Exports	ERS	MacDonald, Steve	202-219-0822
Agricultural Exports	ERS	Warden, Thomas	202-219-0822
Agricultural Imports	ERS	MacDonald, Steve	202-219-0822
Agricultural Imports	ERS	Warden, Thomas	202-219-0822
Apparel Price Indexes, International	BLS	Frumkin, Rob	202-272-5028
Asia, General Information	Census	Hobbs, Frank	301-763-4221
Balance of Payments	BEA	Bach, Christopher	202-523-0620
Balance of Payments, Annual Data	BEA	Fouch, Gregory	202-523-0547
Balance of Payments, Current Account Estimates	BEA	Kealy, Walter	202-523-0625
Balance of Payments, Current Analysis	BEA	DiLullo, Anthony	202-523-0621
Balance of Payments, Quarterly Data	BEA	Fouch, Gregory	202-523-0547
Balance of Payments, Special Analysis	BEA	Lawson, Ann	202-523-0628
Canada, Agricultural Data	ERS	Simone, Mark	202-219-0610
Capital Expenditures, U.S. Foreign Affiliates	BEA	Fahim-Nader, M.	202-523-0640
Capital Transactions, Private	BEA	Scholl, Russell	202-523-0603
China, Agricultural Data	ERS	Tuan, Francis	202-219-0626
China, People's Republic	Census	Banister, Judith	301-763-4012
CIS (USSR), Agricultural Data	ERS	Zeimetz, Kathryn	202-219-0624
Commodity Programs and Policies, World Data	ERS	Dixit, Praveen	202-219-0632
Compensation Cost in Foreign Countries	BLS	Capdevielle, P.	202-523-9292
Cotton, Prices and Other Economic Data	ERS	Whitton, Carolyn	202-219-0824
Credit and Financial Markets, Agriculture	ERS	Baxter, Tim	202-219-0706
Developing Countries, Agricultural Data	ERS	Mathia, Gene	202-219-0680
Direct Investment Abroad, U.S Annual BoPs	BEA	New, Mark	202-523-0612
Direct Investment Abroad, U.S., Analysis	BEA	Lowe, Jeffrey	202-523-0649
Direct Investment Abroad, U.S. Annual Data	BEA	Walker, Patricia	202-523-0612
Direct Investment Abroad, U.S. Benchmark Data	BEA	Walker, Patricia	202-523-0612
Direct Investment Abroad, U.S. Quarterly BoPs	BEA	New, Mark	202-523-0612
East Asia, Agricultural Data	ERS	Coyle, William	202-219-0610
Eastern Europe, Agricultural Data	ERS	Gray, Kenneth	202-219-0621
Eastern Europe, Agricultural Data	ERS	Koopman, Robert	202-219-0621
Eastern Europe, Agricultural Data	ERS	Cochrane, Nancy	202-219-0621
Economic Indicators, Foreign Countries	BLS	Neef, Arthur	202-523-9291

Subject	Source	Data Contact	Telephone
Energy Price Indexes, International	BLS	Fischer, Ilene	202-272-5027
Europe, General Information	Census	Baldwin, Godfrey	301-763-4022
Exports from Manufacturing Establishments	Census	Goldhirsch, Bruce	301-763-1503
Exports, Net, U.S.	BEA	Ehemann, Chris	202-523-0669
Farm Output	ERS	Urban, Francis	202-219-0717
Farm Productivity, World	ERS	Urban, Francis	202-219-0717
Feed Grains, Prices and Economic Data	ERS	Riley, Peter	202-219-0824
Food Demand and Expenditures	ERS	Stallings, Dave	202-219-0708
Food Grains, Prices and Economic Data	ERS	Schwartz, Sara	202-219-0824
Food Policy	ERS	Lynch, Loretta	202-219-0689
Food Price Indexes, International	BLS	Frumkin, Rob	202-272-5028
Foreign Countries, Economic Indicators	BLS	Neef, Arthur	202-523-9291
Foreign Direct Investment (FDI) in U.S.	BEA	Barker, Betty	202-523-0659
Foreign Direct Investment, Annual Survey	BEA	Bomkamp, James	202-523-0559
Foreign Direct Investment, Annual Survey	BEA	Howenstine, Ned	202-523-0650
Foreign Direct Investment, Benchmark Data	BEA	Howenstine, Ned	202-523-0650
Foreign Direct Investment, Benchmark Data	BEA	Bomkamp, James	202-523-0559
Foreign Land Ownership	ERS	DeBraal, Peter	202-219-0425
Foreign Trade, Automated Data Reporting	Census	Brown, Anita	301-763-7700
Foreign Trade Classifications System	Census	Barna, John	301-763-7766
Foreign Trade Commodity Analysis	Census	DeCiccio, Paul	301-763-5200
Foreign Trade, Current Systems Programming	Census	Tormey, George	301-763-7750
Foreign Trade Data Analysis and Planning	Census	Farrell, Michael	301-763-2700
Foreign Trade in Chemicals and Sundries	Census	Norfolk, Irving	301-763-5186
Foreign Trade in Food, Animals, Wood Products	Census	Still, Gloria	301-763-5211
Foreign Trade in Machinery and Vehicles	Census	Herrick, Paul	301-763-5200
Foreign Trade in Minerals and Metals	Census	Staff	301-763-5150
Foreign Trade in Textiles	Census	Wysocki, Adam	301-763-5138
Foreign Trade Information	Census	Kaufman, Milton	301-763-5940
Foreign Trade Procedures and Processing	Census	Robeson, Dwight	301-763-4340
Foreign Trade Programs	Census	Kotwas, Gerald	301-763-5333
Foreign Trade Regulations	Census	Blyweiss, Hal	301-763-5310
Foreign Trade, Special Projects	Census	Becker, Gerald	301-763-7126
Foreign Trade, Transportation	Census	Tague, Norman	301-763-7770
Government Transactions, International	BEA	Thomas, Gregory	202-523-0615
Health Statistics, International	Census	Way, Peter	301-763-4086
Import and Export Price Indexes, U.S.	BLS	Alterman, Bill	202-272-5020
International Data Base	Census	Johnson, Peter	301-763-4811
Labor Costs in Foreign Countries	BLS	Neef, Arthur	202-523-9291
Labor Force in Foreign Countries	BLS	Sorrentino, C.	202-523-9301
Latin America	Census	Hobbs, Frank	301-763-4221

International and Foreign Trade Data — XI. Business Data Directory

Subject	Source	Data Contact	Telephone
Latin America, Agricultural Data	ERS	Link, John	202-219-0660
Livestock, Prices and Economic Data	ERS	Shagam, Shayle	202-219-0767
Livestock, Prices and Economic Data	ERS	Bailey, Linda	202-219-1286
Machinery Price Indexes, International	BLS	Costello, Brian	202-272-5034
Macroeconomic Conditions	ERS	Baxter, Tim	202-219-0706
Manufacturing Establishments, Exports	Census	Goldhirsch, Bruce	301-763-1503
Merchandise Trade, International	BEA	Murad, Howard	202-523-0668
Middle East, Agricultural Data	ERS	Kurtzig, Michael	202-219-0680
Military Sales, Foreign	BEA	McCormick, Bill	202-523-0619
Multinational Corporations, Analysis	BEA	Mataloni, Ray	202-523-3451
Natural Resource Policy, World	ERS	Urban, Francis	202-219-0717
Oceana and North America, General Information	Census	Hobbs, Frank	301-763-4221
Pacific Rim, Agricultural Data	ERS	Coyle, William	202-219-0610
Peanuts, Prices and Economic Data	ERS	McCormick, Ian	202-219-0840
Population, World Data	ERS	Goode, Charles	202-219-0705
Poultry, Prices and Economic Data	ERS	Witucki, Larry	202-219-0766
Price Index Methodology, International	BLS	Alterman, Bill	202-272-5020
Price Index Revisions, International	BLS	Reut, Katrina	202-272-5025
Price Indexes, International	BLS	Reut, Katrina	202-272-5025
Prices in Foreign Countries	BLS	Pettis, Maureen	202-523-9301
Productivity in Foreign Countries	BLS	Neef, Arthur	202-523-9291
Raw Materials Price Indexes, International	BLS	Frumkin, Rob	202-272-5028
Sales, Foreign Military	BEA	McCormick, Bill	202-523-0619
Services Price Indexes, International	BLS	Fischer, Ilene	202-272-5027
Services Transactions, U.S. Analysis	BEA	Whichard, Obie	202-523-0646
Services Transactions, U.S. Analysis	BEA	DiLullo, Anthony	202-523-0621
Services Transactions, U.S. Annual Surveys	BEA	Emond, Chris	202-523-0632
Services Transactions, U.S. Benchmark Data	BEA	Kozlow, Ralph	202-523-0632
Services, U.S. Transactions, Analysis	BEA	Whichard, Obie	202-523-0646
Services, U.S. Transactions, Analysis	BEA	DiLullo, Anthony	202-523-0621
Services, U.S. Transactions, Benchmark Survey	BEA	Kozlow, Ralph	202-523-0632
Shipper's Export Information	Census	Blyweiss, Hal	301-763-5310
South Asia, Agricultural Data	ERS	Landes, Rip	202-219-0664
Soviet Union, General Information	Census	Kostinsky, Barry	301-763-4022
Soybeans, World Prices and Economic Data	ERS	Morgan, Nancy	202-219-0826
Sunflowers, World Prices and Economic Data	ERS	Morgan, Nancy	202-219-0826
Trade Data Services	Census	Staff	301-763-5140
Trade Data Services	Census	Mearkle, Haydn	301-763-7754
Transportation, International	BEA	Watts, Patricia	202-523-0611
Travel, International	BEA	Bolyard, Joan	202-523-0609
Unemployment in Foreign Countries	BLS	Sorrentino, C.	202-523-9301

Subject	Source	Data Contact	Telephone
Western Europe, Agricultural Data	ERS	Coyle, William	202-219-0610
Women in Developing Countries	Census	Jamison, Ellen	301-763-4086

Personal Income and Related Data

Subject	Source	Data Contact	Telephone
Administrative Salaries	BLS	Smith, William	202-523-1570
Area Wage Surveys	BLS	Buckley, John	202-523-1763
Average Pay Data (Unemployment Insurance Data)	BLS	Bush, Joseph	202-523-1158
Capital Accumulation Benefits	BLS	Houff, James	202-523-8791
Clerical Salaries	BLS	Smith, William	202-523-1570
Disability Benefits and Paid Leave	BLS	Wiatrowski, Bill	202-523-8791
Disposable Personal Income	BEA	Cypert, Pauline	202-523-0832
Disposable Personal Income, State	BEA	Brown, Robert	202-254-6632
Dividends	BEA	Petrick, Kenneth	202-523-0888
Dividends, Subnational Estimates	BEA	Jolley, Charles	202-254-6637
Earnings, Annual, CPS	BLS	Mellor, Earl	202-523-1371
Earnings, Weekly, CPS	BLS	Mellor, Earl	202-523-1371
Employee Compensation	BEA	Sensenig, Arthur	202-523-0809
Farm Household Income	ERS	Ahearn, Mary	202-219-0807
Farm Income	ERS	McElroy, Bob	202-219-0800
Farm Income	BEA	Smith, George	202-523-0821
Farm Income	ERS	Strickland, Roger	202-219-0804
Farm Proprietors' Income, Subnational Data	BEA	Zavrel, James	202-254-6638
Farm Wages	ERS	Duffield, James	202-219-0033
Farm Wages	NASS	Kurtz, Tom	202-690-3228
Farm Wages	ERS	Oliveira, Victor	202-219-0033
Household Wealth	Census	Lamas, Enrique	301-763-8578
Income Information, Decennial Censuses	Census	Downs, Bill	301-763-8553
Income Statistics	Census	Staff	301-763-8576
Income, Interest	BEA	Weadock, Teresa	202-523-0833
Income, National	BEA	Seskin, Eugene	202-523-0848
Income, Nonfarm Proprietors'	BEA	Abney, Willie	202-523-0811
Income, Personal	BEA	Cypert, Pauline	202-523-0832
Income, Rental	BEA	Smith, George	202-523-0821
Industry Earnings, Monthly Survey (CPS)	BLS	Parks, William	202-523-1959
Industry Wage Surveys	BLS	Williams, Harry	202-523-1667
Interest Income	BEA	Weadock, Teresa	202-523-0833
Interest Payments	BEA	Weadock, Teresa	202-523-0833
Interest, Subnational Estimates	BEA	Jolley, Charles	202-254-6637
Minimum Wage Data	BLS	Haugen, Steve	202-523-1944
National Income	BEA	Seskin, Eugene	202-523-0848

Subject	Source	Data Contact	Telephone
Nonfarm Proprietors' Income, Subnational	BEA	Levine, Bruce	202-254-6634
Payrolls, County Business Patterns	Census	Decker, Zigmund	301-763-5430
Payrolls, Monthly Survey (BLS-790)	BLS	Seifert, Mary Lee	202-523-1172
Personal Income Data	BEA	Cypert, Pauline	202-523-0832
Personal Income, Counties	BEA	Hazen, Linnea	202-254-6642
Personal Income Methodology	BEA	Bailey, Wallace	202-254-6635
Personal Income, Metropolitan Areas	BEA	Hazen, Linnea	202-254-6642
Personal Income, State	BEA	Hazen, Linnea	202-254-6642
Personal Income, Subnational Data	BEA	Staff	202-254-6630
Professional Salaries	BLS	Smith, William	202-523-1570
Proprietors' Income, Nonfarm	BEA	Abney, Willie	202-523-0811
Real Earnings Data, Current	BLS	Ulmer, Mark	202-523-1172
Rental Income	BEA	Smith, George	202-523-0821
Rental Income, Subnational Estimates	BEA	Jolley, Charles	202-254-6637
Retirement Benefits	BLS	Houff, James	202-523-8791
Salaries, Clerical	BLS	Smith, William	202-523-1570
Salaries, Professional	BLS	Smith, William	202-523-1570
Salaries, Technical	BLS	Smith, William	202-523-1570
Service Contract Act Wage Surveys	BLS	Van Giezen, Bob	202-523-1536
SIPP, General Information	Census	Staff	301-763-2764
SIPP Survey Income Program Participation	Census	Bowie, Chester	301-763-2764
State Personal Income	BEA	Hazen, Linnea	202-254-6642
State Quarterly Personal Income	BEA	Whiston, Isabelle	202-254-6672
Supplemental Data System, Workers Compensation	BLS	Anderson, John	202-501-6463
Survey of Income and Program Participation-SIPP	Census	Kasprzyk, Daniel	301-763-8328
Survey of Income and Program Participation-SIPP	Census	Staff	301-763-2764
Technical Salaries	BLS	Smith, William	202-523-1570
Transfer Payments, Subnational Data	BEA	Levine, Bruce	202-254-6634
Wage Chronologies	BLS	Cimini, Michael	202-523-1320
Wages and Salaries Data	BEA	Sensenig, Arthur	202-523-0809
Wages and Salaries, Subnational Data	BEA	Carnevale, Sharon	202-523-0945
Wages by Industry, State and Area	BLS	Bush, Joseph	202-523-1158
Wages, Current Developments	BLS	Cimini, Michael	202-523-1320
Wages, Quarterly Data (ES-202 Reports)	BLS	Bush, Joseph	202-523-1158
Wages, State and Local Governments	BLS	Field, Charles	202-523-1570
Weekly and Annual Earnings Data, CPS	BLS	Mellor, Earl	202-523-1371

Population and Other Demographic Data

Subject	Source	Data Contact	Telephone
Age and Sex (States and Counties)	Census	Staff	301-763-5072
Aging Population	Census	Goldstein, Arnold	301-763-7883
Apportionment	Census	Speaker, Robert	301-763-7962
Births and Fertility	Census	Bachu, Amara	301-763-5303
Births and Fertility	Census	O'Connell, Martin	301-763-5303
Centers of Population	Census	Hirschfeld, Don	301-763-5720
Child Care Data	Census	O'Connell, Martin	301-763-5303
Child Care Data	Census	Bachu, Amara	301-763-5303
Citizenship Data	Census	Staff	301-763-7955
Commuting, Journey to Work	Census	Swieczkowski, G.	301-763-3850
Commuting, Journey to Work	Census	Salopek, Phil	301-763-3850
Commuting, Means of Transportation	Census	Salopek, Phil	301-763-3850
Commuting, Means of Transportation	Census	Boertlein, Celia	301-763-3850
Commuting, Place of Work	Census	Boertlein, Celia	301-763-3850
Commuting, Place of Work	Census	Salopek, Phil	301-763-3850
Confidentiality and Privacy Issues	Census	Gates, Jerry	301-763-5062
Crime Statistics	Census	McGinn, Larry	301-763-1735
Decennial Census, 1990 Count Information	Census	Staff	301-763-5002
Decennial Census, 1990 Count Questions	Census	Kobilarcik, Ed	301-763-4894
Decennial Census, General Information	Census	Paez, Al	301-763-4251
Disability Estimates	Census	McNeil, Jack	301-763-8300
Education Statistics	Census	Staff	301-763-1154
Emigration	Census	Woodrow, Karen	301-763-5590
Ethnic Population Data	Census	Staff	301-763-7955
Ethnic Population Data, Excluding Hispanics	Census	Staff	301-763-7955
Families	Census	Staff	301-763-7987
Farm Population	Census	Dahmann, Don	301-763-5158
Fertility and Births	Census	O'Connell, Martin	301-763-5303
Fertility and Births	Census	Bachu, Amara	301-763-5303
Foreign Born Population	Census	Staff	301-763-7955
Health Surveys	Census	Mangold, Robert	301-763-5508
Hispanic Population Data	Census	Staff	301-763-7955
Homeless Population	Census	Taeuber, Cynthia	301-763-7883
Household Estimates for States and Counties	Census	Staff	301-763-5221
Households and Families	Census	Staff	301-763-7987
Immigration, Legal and Undocumented	Census	Woodrow, Karen	301-763-5590

Part Two: The DATAPHONER *Population and Other Demographic Data*

Subject	Source	Data Contact	Telephone
Language Information	Census	Staff	301-763-1154
Living Arrangements	Census	Saluter, Arlene	301-763-7987
Longitudinal Surveys	Census	Dopkowski, Ron	301-763-2767
Marital Status	Census	Saluter, Arlene	301-763-7987
Migration Data	Census	DeAre, Diana	301-763-3850
National Age and Sex Data	Census	Staff	301-763-7950
National Population Estimates	Census	Staff	301-763-7950
National Population Projections	Census	Staff	301-763-7950
Outlying Area Information	Census	Staff	301-763-2903
Place of Birth Statistics	Census	Hansen, Kristin	301-763-3850
Population Censuses	Census	Staff	301-763-5020
Population Censuses, Special	Census	Dopkowski, Ron	301-763-2767
Population Data, Special Tabulations	Census	Cowan, Rosemarie	301-763-7947
Population Estimates	Census	Tucker, Ronald	301-763-2773
Population Information	Census	Staff	301-763-5002
Population of Outlying Areas	Census	Levin, Michael	301-763-5134
Population Survey, Current	Census	Tucker, Robert	301-763-2773
Population Surveys, Special	Census	Dopkowski, Ron	301-763-2767
Poverty Statistics	Census	Staff	301-763-8578
Prisoner Surveys	Census	McGinn, Larry	301-763-1735
Projections, National Population	Census	Staff	301-763-7950
Projections, State Population	Census	Staff	301-763-1902
Race Statistics	Census	Staff	301-763-7572
Race Statistics	Census	Staff	301-763-2607
Sampling Methods, Decennial Census	Census	Woltman, Henry	301-763-5987
School District Data	Census	Ingold, Jane	301-763-3476
States and Outlying Areas Estimates	Census	Staff	301-763-5072
Statistical Research for Demographic Programs	Census	Ernst, Lawernce	301-763-7880
Survey Operations	Census	Speight, G.	301-763-7783
Unemployment Statistics	Census	Jones, Selwyn	301-763-8574
Veterans Status	Census	Palumbo, Thomas	301-763-8574
Veterans Status	Census	Jones, Selwyn	301-763-8574
Wealth, Households	Census	Lamas, Enrique	301-763-8578
Zip Code Demographic Data	Census	Staff	301-763-4100

Price and Price Index Data

Subject	Source	Data Contact	Telephone
Chain Price Indexes	BEA	Herman, Shelby	202-523-0828
Chemical Producers Price Index	BLS	Gaddie, Robert	202-272-5210
Computer Price Index	BEA	Won, Gregory	202-523-5421
Construction, Producer Price Index	BLS	Davis, Wanda	202-272-5127
Consumer Expenditure Survey Data and Tapes	BLS	Passero, William	202-272-5060
Consumer Expenditure Survey Data and Tapes	BLS	Dietz, Richard	202-272-5156
Consumer Expenditure Surveys	BLS	Jacobs, Eva	202-272-5156
Consumer Price Index	BLS	Jackman, Patrick	202-272-5160
Consumer Price Index, Food	ERS	Parlett, Ralph	202-219-0870
Consumer Price Index, Food	ERS	Dunham, Denis	202-219-0870
Electric Machinery Producer Price Index	BLS	Sinclair, James	202-272-5052
Energy Producers Price Index	BLS	Gaddie, Robert	202-272-5127
Export and Import Price Indexes, U.S.	BLS	Alterman, William	202-272-5020
Family Budget	BLS	Rogers, John	202-272-5060
Farm Prices, Parity and Indexes	NASS	Milton, Bob	202-720-3570
Farm Prices, Parity Paid	NASS	Kleweno, Doug	202-720-4214
Farm Prices, Parity Received	NASS	Buche, John	202-720-5446
Fixed-Weighted Indexes	BEA	Herman, Shelby	202-523-0828
Food, Average Monthly Retail Prices	BLS	Cook, William	202-272-5173
Food Prices and Consumer Price Index	ERS	Parlett, Ralph	202-219-0870
Food Prices and Consumer Price Index	ERS	Dunham, Denis	202-219-0870
Food Producers Price Index	BLS	Gaddie, Robert	202-272-5210
Foreign Countries, Prices	BLS	Pettis, Maureen	202-523-9301
Forestry Producer Price Index	BLS	Davis, Wanda	202-272-5127
Fruits and Vegetables, Price Spreads	ERS	Parrow, Joan	202-219-0883
Fuel Price Indexes, Monthly	BLS	Adkins, Robert	202-272-5177
Honey, Price Data	NASS	Schuchardt, R.	202-690-3236
International Apparel Price Indexes	BLS	Frumkin, Rob	202-272-5028
International Energy Price Indexes	BLS	Fischer, Ilene	202-272-5027
International Food Price Indexes	BLS	Frumkin, Rob	202-272-5028
International Machinery Price Indexes	BLS	Costello, Brian	202-272-5034
International Price Index Methodology	BLS	Alterman, Bill	202-272-5020
International Price Index Revisions	BLS	Reut, Katerina	202-272-5025
International Raw Materials Price Indexes	BLS	Frumkin, Rob	202-272-5028
International Services Price Indexes	BLS	Fischer, Ilene	202-272-5027
Leather, Producer Price Index	BLS	Paik, Soon	202-272-5127

Subject	Source	Data Contact	Telephone
Meat, Price Spreads	ERS	Duewer, Larry	202-219-0712
Metals, Producer Price Index	BLS	Kazanowski, Ed	202-272-5204
Motor Fuels, Average Retail Prices	BLS	Chelena, Joseph	202-272-5080
New Residential Construction Price Index	Census	Berman, Steve	301-763-7842
Non-Electric Machinery Producers Price Index	BLS	Lavish, Kenneth	202-272-5115
Personal Consumption Expenditures Data	BEA	McCully, Clint	202-523-0819
Price Indexes for Motor Fuels	BLS	Chelena, Joseph	202-272-5080
Price Indexes, Service Industry	BLS	Weeden, George	202-272-5130
Prices Indexes, Farm	NASS	Milton, Bob	202-720-3570
Producer Price Index, Analysis and Data	BLS	Howell, Craig	202-272-5113
Producer Price Index Methodology	BLS	Rosenberg, Elliott	202-272-5118
Producer Price Indexes	BLS	Tibbetts, Thomas	202-272-5110
Retail Food Prices, Monthly Estimates	BLS	Rabil, Floyd	202-272-5173
Service Industry Price Indexes	BLS	Weeden, George	202-272-5130
Services, Producer Price Index	BLS	Gerduk, Irwin	202-523-5130
Statistical Methods	BLS	Hedges, Brian	202-272-2195
Textiles, Producer Price Index	BLS	Paik, Soon	202-272-5127
Transportation Equipment, Producer Price Index	BLS	Sinclair, James	202-272-5052
Transportation Price Indexes	BLS	Royce, John	202-272-5131
Utility Price Indexes, Monthly	BLS	Adkins, Robert	202-272-5177

Rural Development Data

Subject	Source	Data Contact	Telephone
Agriculture and Community Linkages	ERS	Hines, Fred	202-219-0525
Business and Industry	ERS	Bernat, Andrew	202-219-0540
Community Development	ERS	Sears, David	202-219-0544
Credit and Finance Markets, Rural Data	ERS	Sullivan, Pat	202-219-0719
Credit and Finance Markets, Rural Data	ERS	Rossi, Cliff	202-219-0892
Employment, Rural Areas	ERS	Swaim, Paul	202-219-0552
Employment, Rural Areas	ERS	Parker, Tim	202-219-0541
Income and Poverty, Rural Statistics	ERS	Hoppe, Robert	202-219-0547
Local Government Finances, Rural Data	ERS	Jansen, Anicca	202-219-0542
Local Government Finances, Rural Data	ERS	Reeder, Richard	202-219-0542
Rural Development Information	ERS	McGranaham, D.	202-219-0532
Rural Development Information	ERS	Mazie, Sara	202-219-0530

Statistical Concepts and Methods

Subject	Source	Data Contact	Telephone
Area Measurement	Census	Hirschfeld, Don	301-763-5720
Areas of Economic Censuses	Census	Staff	301-763-4667
BEA Economic Areas	BEA	Trott, Edward	202-523-0973
Business Cycle, Methodology	BEA	Beckman, Barry	202-523-0800
Census Cartographic Operations	Census	Staff	301-763-3973
Census Geographic Concepts	Census	Staff	301-763-5720
Census Tract Boundaries	Census	Miller, Cathy	301-763-3827
Census Tract Codes	Census	Miller, Cathy	301-763-3827
Census Tract Deliniation	Census	Miller, Cathy	301-763-3827
Census Tracts Address Locations	Census	Swapshur, Ernie	301-763-5720
Commodity Classification Information	Census	Venning, Alvin	301-763-1935
Decennial Census Program Design	Census	Berman, Patricia	301-763-7094
Demographic Analysis	Census	Robinson, Gregg	301-763-5590
Employment, Subnational Methodology	BEA	Bailey, Wallace	202-254-6635
Foreign Trade, Methodology	Census	Oberg, Diane	301-763-5709
Foreign Trade, Methodology and Quality Assurance	Census	Walter, Bruce	301-763-7020
Foreign Trade, Methods and Research	Census	Puzzilla, Kathleen	301-763-7760
Geographic Areas of Economic Censuses	Census	Staff	301-763-4667
Geographic Concepts, Census	Census	Staff	301-763-5720
Industrial Classification, Methodology	BLS	Pinkos, John	202-523-1636
Industry Classification Information	Census	Venning, Alvin	301-763-1935
Input-Output Tables, Methodology	BEA	Maley, Leo	202-523-0683
Personal Income, Subnational Methodology	BEA	Bailey, Wallace	202-254-6635
Post-Enumerations Surveys	Census	Hogan, Howard	301-763-1794
Price and Index Number Research	BLS	Zieschang, Kim	202-272-5096
Producer Price Index Revisions	BLS	Tibbetts, Thomas	202-272-5110
Productivity Research	BLS	Harper, Michael	202-523-6010
Quarterly Financial Report	Census	Lee, Ronald	301-763-4270
Quarterly Financial Report, Classification	Census	Hartman, Frank	301-763-4274
Regional Income and Employment Measures	BEA	Bailey, Wallace	202-254-6635
Residence Adjustments to Regional Income	BEA	Zabronsky, Daniel	202-254-6639
Sampling Methods	Census	Waite, Preston	301-763-2672
Seasonal Adjustment Methodology	BLS	McIntire, Robert	202-523-1776
SIPP Statistical Methods	Census	Singh, Raj	301-763-7944
Standard Occupational Classification System	BLS	McElroy, Michael	202-523-1684
State Boundary Certification	Census	Stewart, Louisa	301-763-3827

Subject	Source	Data Contact	Telephone
State Shift-Share Analysis, Methodology	BEA	Kort, John	202-523-0946
Statistical Areas	Census	Staff	301-763-3827
Statistical Research for Economic Programs	Census	Monsour, Nash	301-763-5702
UN and OECD System of National Accounts	BEA	Honsa, Jeanette	202-523-0835

Unemployment Data

Subject	Source	Data Contact	Telephone
Claimant Data, Unemployment Insurance Data	BLS	Brown, Sharon	202-523-1038
Discouraged Workers	BLS	Hamel, Harvey	202-523-1371
Displaced Workers	BLS	Horvath, Francis	202-523-1371
Economic Hardship Data	BLS	Klein, Bruce	202-523-1371
Mass Layoff Statistics	BLS	Siegal, Lewis	202-523-1807
Occupational Unemployment Survey (OESS)	BLS	Cromartie, Stella	202-523-1371
Plant Closing Statistics	BLS	Siegal, Lewis	202-523-1807
Unemployment Insurance Claimants	BLS	Terwilliger, Y.	202-523-1002
Unemployment Statistics	Census	Palumbo, Thomas	301-763-8574

Working Conditions and Compensation Data

Subject	Source	Data Contact	Telephone
Benefits for Part-Time Workers	HA	Schmidt, C.	708-295-5000
Benefits, General Information	EBRI	Staff	202-659-0670
Benefits in a Specific Area	BLS	Buckley, John	202-523-1763
Benefits in a Specific Industry	BLS	Williams, Harry	202-523-1667
Benefits in Small Establishments	HA	Schmidt, C.	708-295-5000
Benefits in Small Establishments	EBRI	Staff	202-659-0670
Child Care	WB	Staff	202-523-6652
Collective Bargaining Agreements, Analysis	BLS	Cimini, Michael	202-523-1320
Collective Bargaining, Public File	BLS	Cimini, Michael	202-523-1320
Collective Bargaining Settlements, Major	BLS	Devine, Janice	202-523-1308
Cost of Providing Health Care to Employees	HCBS	Blake, Pamela	609-520-2289
Disability Benefits	BLS	Morton, John	202-523-8791
Disability Benefits and Paid Leave	BLS	Wiatrowski, Bill	202-523-8791
Employee Absences from Work	DHHS	Staff	301-436-8500
Employee Absences from Work	BLS	Meisenheimer, Joe	202-523-1944
Employee Associations, Membership Data	BLS	Cimini, Michael	202-523-1320
Employee Benefit Plans	BEA	Sensenig, Arthur	202-523-0809
Employee Benefits Survey	BLS	Staff	202-523-9445
Employee Compensation	BEA	Sensenig, Arthur	202-523-0809
Employees Rights to Benefits	PWBA	Small, Richard	202-523-8776
Enforcement of Pension Laws	PWBA	Small, Richard	202-523-8776
ERISA	PWBA	Beller, Dan	202-523-9505
Financial Information	PWBA	Beller, Dan	202-523-9505
Flexible Benefits Programs	HCBS	Higgins, Foster	609-520-2441
Health Care Cost to Employees	HCFA	McPhillips, R.	301-597-3934
Health Care Costs in Business	SBA	Lichtenstein, Jules	202-653-6365
Health Information	HIAA	White, Don	202-223-7782
Health Information	HIAA	Musco, Thomas	202-223-7863
Health Information	HIAA	Minor, Al	202-223-7845
Health Insurance Benefits	BLS	Blostin, Allan	202-523-8791
Health Insurance, Americans Covered	PWBA	Beller, Dan	202-523-9505
Health Insurance, Americans Not Covered	HCFA	Staff	301-597-3934
Health Studies and Special Projects	BLS	Hilaski, Harvey	202-272-3459
Illness, Characteristics	BLS	Anderson, John	202-501-6463
Illnesses in Industry, Estimates	BLS	Jackson, Ethel	202-501-6470
Illnesses in Industry, Incidence Rates	BLS	Jackson, Ethel	202-501-6470

Subject	Source	Data Contact	Telephone
Injuries, Characteristics	BLS	Anderson, John	202-501-6463
Injuries in Industry, Estimates	BLS	Jackson, Ethel	202-501-6470
Injuries in Industry, Incidence Rates	BLS	Jackson, Ethel	202-501-6470
Life Insurance Benefits	BLS	Blostin, Allan	202-523-8791
Maternity Leave	WLDF	Staff	202-887-0364
Occupational Injuries, Fatal	BLS	Toscano, Guy	202-501-6459
Occupational Injury Data, Annual Survey	BLS	Jackson, Ethel	202-501-6470
OSHA Recordkeeping Requirements	BLS	Whitmore, Robert	202-272-3462
Paid Leave	BLS	Morton, John	202-523-8791
Parental Leave	WLDF	Staff	202-887-0364
Pay and Benefit Differentials, Union and Non-Union	BLS	Shelley, Wayne	202-523-1165
Pension Plans, Number of Americans Covered	PWBA	Beller, Dan	202-523-9505
Pensions, General Information	PWBA	Beller, Dan	202-523-9505
Profit Sharing Plans	PSRF	Bell, John	312-372-3416
Retirees Health Coverage	HCBS	Higgins, Foster	609-520-2441
Retirement Benefits	BLS	Houff, James	202-523-8791
Sick Leave	DHHS	Staff	301-436-8500
Sick Leave	BLS	Meisenheimer, Joe	202-523-1944
Small Business Employee Benefits	NFIB	Dennis, William	202-554-9000
Substance Abuse Data	NIAAA	Montague, Barry	301-443-3864
Union and Non-Union Pay and Benefit Differentials	BLS	Shelley, Wayne	202-523-1165
Union Statistics	BLS	Cimini, Michael	202-523-1320
Work Injury Reports Survey	BLS	Carlson, Norma	202-501-7570
Work Stoppages	BLS	Cimini, Michael	202-523-1320
Work-Life Estimates	BLS	Horvath, Francis	202-523-1371

XI. Business Data Directory

Section II. State and Regional Data Centers

This section guides you to the federal government's state and regional data centers. All of the major economic Data Factories have regional affiliates—the Bureau of the Census (Census), the Bureau of Economic Analysis (BEA), the Bureau of Labor Statistics (BLS) and the National Agricultural Statistical Service (NASS). These affiliates are located in all 50 states and the District of Columbia. BEA, BLS and Census data are also available in Puerto Rico, the Virgin Islands and Guam.

While BEA maintains an elaborate network of university and state government affiliates to distribute its economic data, the Census Bureau maintains the most ambitious and wide reaching network for the redistribution of Census data. The Census Bureau's most recent additions to its regional data network are the **Business and Industry Data Centers** located in 16 states. These centers are set up to channel economic and marketing data from the Census Bureau to local business and industry. Agriculture (NASS) maintains State Statisticians' Offices in 45 states for the collection and distribution of local agricultural data. BLS maintains eight regional offices to distribute labor-related data.

The state and regional data centers are listed below in the following order:

The Census Bureau: State Data Centers

The Census Bureau: Business and Industry Data Centers

The Bureau of Economic Analysis: State User Groups

The Bureau of Labor Statistics: Regional Offices

The National Agricultural Statistics Service: State Statisticians' Offices

The Census Bureau: State Data Centers (SDCs)

The Census Bureau has a network of lead agencies in its State Data Centers (SDCs) program. Data from the Bureau of Economic Analysis (BEA) is also available at many of the Census Bureau's State Data Centers.

Under a cooperative agreement with the Census Bureau, each state maintains a state data center to redistribute census data within the state. Each state has a lead data center and a network of affiliates within the state (city and regional planning agencies, local libraries, etc.). Lead centers have sets of Census data and provide training on their use. Affiliates serve as statistical resource centers in their communities.

If you have any general questions about the Census Bureau's products and services, you can get answers from the information specialists in the Census Bureau's 12 regional offices. These specialists know the local situation and can usually tell you whom to contact for other sources of economic or related data. Keep in mind that the information specialists at the 12 regional offices can answer only general questions about data. If you have a specific question or need more detailed information, contact the data experts listed in the **DATAPHONER**.

For a fee, the Census Bureau's state centers provide the following services:

- Download computer tape files to microcomputer diskettes.
- State economic profiles with information from the Census Bureau and other government and private sources.
- Special extracts from large Census Bureau tape files.
- Online data sources.
- Marketing research.
- Guides to local data sources.
- Small-area profiles and studies on issues of local importance.
- Maps for special areas.
- Access to reference libraries.

The Census Bureau State Data Centers
(Lead Data Centers)

ALABAMA
Center for Business and
Economic Research
University of Alabama
Tuscaloosa, AL 35487
(205) 348-6191

ALASKA
Research and Analysis
Department of Labor
Juneau, AK 99802
(907) 465-4500

ARIZONA
Arizona Department of
Economic Security
Phoenix, AZ 85005
(602) 542-5984

ARKANSAS
University of Arkansas at
Little Rock
Little Rock, AR 72204
(501) 569-8530

CALIFORNIA
Department of Finance
Sacramento, CA 95814
(916) 322-4651

COLORADO
Colorado Department of
Local Affairs
Denver, CO 80203
(303) 866-2156

CONNECTICUT
Connecticut Office of
Policy and Management
Hartford, CT 06106
(203) 566-8285

DELAWARE
Delaware Development Office
Dover, DE 19903
(302) 739-4271

DISTRICT OF COLUMBIA
Mayor's Office of Planning
Washington, DC 20004
(202) 727-6533

FLORIDA
Executive Office of the Governor
Tallahassee, FL 32399
(904) 487-2114

GEORGIA
Office of Planning and Budget
Atlanta, GA 30334
(404) 656-0911

GUAM
Guam Department of Commerce
Tamuning, Guam 96911
(671) 646-5841

HAWAII
State Department of Business and
Economic Development
Honolulu, HI 96804
(808) 548-3082

IDAHO
Idaho Department of Commerce
Boise, ID 83720
(208) 334-2470

ILLINOIS
Illinois Bureau of the Budget
Springfield, IL 62706
(217) 782-1381

INDIANA
Indiana State Data Center
Indiana State Library
Indianapolis, IN 46204
(317) 232-3733

IOWA
State Library of Iowa
Des Moines, IA 50319
(515) 281-4350

KANSAS
State Library
Topeka, KS 66612
(913) 296-3296

KENTUCKY
Urban Research Institute
University of Louisville
Louisville, KY 40292
(502) 588-7990

LOUISIANA
Office of Planning and Budget
Division of Administration
Baton Rouge, LA 70804
(504) 342-7410

MAINE
Maine Department of Labor
Augusta, ME 04330
(207) 289-2271

MARYLAND
Maryland Department of State Planning
Baltimore, MD 21201
(301) 225-4450

MASSACHUSETTS
Massachusetts Institute for Social and Economic Research
University of Massachusetts
Amherst, MA 01003
(413) 545-3460

MICHIGAN
Department of Management and Budget
Lansing, MI 48909
(517) 373-7910

MINNESOTA
State Demographer's Office
Minnesota State Planning Agency
St. Paul, MN 55155
(612) 297-2360

MISSISSIPPI
Center for Population Studies
University of Mississippi
University, MS 38677
(601) 232-7288

MISSOURI
Missouri State Library
Jefferson City, MO 65102
(314) 751-3615

MONTANA
Montana Department of Commerce
Helena, MT 59620
(406) 444-2896

NEBRASKA
Center for Applied Urban Research
University of Nebraska - Omaha
Omaha, NE 68182
(402) 595-2311

NEVADA
Nevada State Library
Carson City, NV 89710
(702) 885-5160

NEW HAMPSHIRE
Office of State Planning
Concord, NH 03301
(603) 271-2155

NEW JERSEY
New Jersey Department of Labor
Trenton, NJ 08625
(609) 984-2593

NEW MEXICO
Economic Development and
Tourism Department
Sante Fe, NM 87503
(505) 827-0276

NEW YORK
New York Department of Economic
Development
Albany, NY 12245
(518) 474-6005

NORTH CAROLINA
North Carolina Office of State
Budget and Management
Raleigh, NC 27603
(919) 733-7061

NORTH DAKOTA
Department of Agricultural
Economics
North Dakota State University
Fargo, ND 58105
(701) 237-8621

OHIO
Ohio Department of Development
Columbus, OH 43266
(614) 466-2115

OKLAHOMA
Oklahoma Department of Commerce
Oklahoma City, OK 73126
(405) 841-5184

OREGON
Center for Population Research and
Census
Portland State University
Portland, OR 97207
(503) 725-3922

PENNSYLVANIA
Institute of State and Regional
Affairs
Pennsylvania State University
Middletown, PA 17057
(717) 948-6336

PUERTO RICO
Puerto Rico Planning Board
San Juan, PR 00940
(809) 728-4430

RHODE ISLAND
Office of Municipal Affairs
Providence, RI 02908
(401) 277-6493

SOUTH CAROLINA
South Carolina Budget and Control
Board
Columbia, SC 29201
(803) 734-3780

SOUTH DAKOTA
Business Research Bureau
University of South Dakota
Vermillion, SD 57069
(605) 677-5287

TENNESSEE
State Planning Office
Nashville, TN 37219
(615) 741-1676

TEXAS
Texas Department of Commerce
Capital Station
Austin, TX 78711
(512) 472-9667

UTAH
Office of Planning and Budget
Salt Lake City, UT 84114
(801) 538-1036

VERMONT
Office of Policy Research and
Coordination
Montpelier, VT 05602
(802) 828-3326

VIRGIN ISLANDS
Caribbean Research Institute
University of the Virgin Islands
Charlotte Amalie
St. Thomas, VI 00802
(809) 776-9200

VIRGINIA
Virginia Employment Commission
Richmond, VA 23219
(804) 786-8308

WASHINGTON
Office of Financial Management
Olympia, WA 98504
(206) 586-2504

WEST VIRGINIA
Governor's Office of Community
and Industrial Development
Charleston, WV 25305
(304) 348-4010

WISCONSIN
Demographic Service Center
Madison, WI 53707
(608) 266-1927

WYOMING
Department of Administration and
Fiscal Control
Cheyenne, WY 82002
(307) 777-7504

The Census Bureau: Business and Industry Data Centers (BIDCs)

The Census Bureau Business and Industry Data Centers (BIDCs) are a recent outgrowth of the State Data Center Program. They were set up to channel economic and marketing data from the Census Bureau to local businesses and industry. They are there to help entrepreneurs get answers to such questions as:

    Is there a local market for my product?

    Will people buy my product or service? How much money do they have?

    Is the prospective market saturated? If so, are there new areas
        where competition is weaker?

    Will I be able to find and afford a location where suppliers and customers
        can be reached?

Some of the Business and Industry Data Centers can give guidance on local markets. Some of the BIDCs can also explain the mechanics of getting a business started—how to set prices, establish proper records, arrange stock and manage taxes.

For additional information on the Business and Industry Data Centers call the Data User Services Division of the Census Bureau at: **(301) 763-1580**.

The Census Bureau Business and Industry Data Centers (BIDCs)

CONNECTICUT
Connecticut Office of Policy and
Management
Hartford, CT 06106
(203) 566-8285

DELAWARE
Delaware Development Office
Dover, DE 19903
(302) 739-4271

FLORIDA
Executive Office of the Governor
Tallahassee, FL 32399
(904) 487-2114

INDIANA
Indiana Business Research Center
Bloomington, IN 47405
(812) 855-5507

Indiana Business Research Center
Indianapolis, IN 46202
(317) 274-2205

KENTUCKY
Urban Research Institute
University of Louisville
Louisville, KY 40292
(502) 588-7990

MARYLAND
Maryland Department of State
Planning
Baltimore, MD 21201
(301) 225-4450

MASSACHUSETTS
Massachusetts Institute for Social
and Economic Research
University of Massachusetts
Amherst, MA 01003
(413) 545-3460

MINNESOTA
State Demographer's Office
Minnesota State Planning Agency
St. Paul, MN 55155
(612) 297-2360

MONTANA
Montana Department of Commerce
Helena, MT 59620
(406) 444-2896

NEW JERSEY
New Jersey Department of Labor
Trenton, NJ 08625
(609) 984-2593

NEW MEXICO
Bureau of Business and Economic
Research
University of New Mexico
1920 Lomas N.E.
Albuquerque, NM 87131
(505) 277-2216

NORTH CAROLINA
North Carolina Office of State
Budget and Management
Raleigh, NC 27603
(919) 733-7061

PENNSYLVANIA
Institute of State and Regional
Affairs
Pennsylvania State University
Middletown, PA 17057
(717) 948-6336

WASHINGTON
Office of Financial Management
Olympia, WA 98504
(206) 586-2504

WEST VIRGINIA
Center for Economic Research
West Virginia University
Morgantown, WV 26506
(304) 293-5837

WISCONSIN
Applied Population Laboratory
University of Wisconsin
Madison, WI 53706
(608) 262-9526

WYOMING
Department of Administration
and Fiscal Control
Cheyenne, WY 82002
(307) 777-7504

The Bureau of Economic Analysis: State User Groups

The Bureau of Economic Analysis (BEA) was one of the first economic Data Factories to set up a system of state data centers. BEA also distributes income, employment and other local area data series through a group of universities and state agencies. State centers were created by a Congressional directive requiring BEA to provide its regional data to universities and state agencies on a no-cost basis. State users receive a complete set of BEA's regional economic data products for the relevant state, its counties, and its metropolitan areas on the condition that the users will in turn distribute the data within the state.

BEA's state and local personal income, employment and other local area data are made available to data users through its Regional Economic Information System. This was put in place for the maintenance, management and distribution of BEA's regional data base. At the present time, the system contains nearly 40 million separate estimates covering over 3,500 local areas. It expands at the rate of 2 million estimates per year.

BEA produces estimates of total and per capita personal income (and related measures) for states, metropolitan areas and counties. Personal income—wages and salaries by industry; other labor income; proprietors' income; rental income for persons, dividends, and personal interest income; transfer payments; and personal contributions for social insurance—are some of the most widely used measures of regional and local area economic well-being. The wealth of relatively current local-area industrial detail contained in BEA's state, metropolitan and county personal income series has historically served as the basis for the analysis of regional and local area economies.

For more information on BEA's Regional Economic Information System write to:

The Regional Economic Measurement Division
BE-55
Bureau of Economic Analysis
Washington, DC 20230

or call: **(202) 254-6642**.

The Bureau of Economic Analysis State User Groups

ALABAMA

Dr. Semoom Chang
Director
Center for Business and Economic Research
University of South Alabama
Mobile, AL 36688
(205) 460-6156

Mr. Parker Collins
Alabama Department of Economic and Community Affairs
P.O. Box 250347
Montgomery, AL 36125
(205) 284-8700

Mr. Douglas Dyer
Chief
Alabama Department of Industrial Relations
649 Monroe Street
Montgomery, AL 36688
(205) 242-8855

Dr. Mac R. Holmes
Center for Business and Economic Services
Troy State University
Troy, AL 36082
(205) 670-3144

Ms. Deborah Hamilton
Center for Business and Economic Research
University of Alabama
Box 870221
Tuscaloosa, AL 35487
(205) 348-6191

ALASKA

Mr. Lee Gorsuch
Director
Institute of Social and Economic Research
University of Alaska
3211 Providence Drive
Anchorage, AK 99508
(907) 786-7710

Mr. Gregg Erickson
Division of Policy
Pouch AD
Juneau, AK 99811
(907) 465-3568

Ms. Kathryn Lizik
Alaska Department of Labor
Research and Analysis
State Data Center
P.O. Box 25504
Juneau, AK 99801
(907) 465-4500

ARIZONA

Dr. Max Jerrell
College of Business Administration
Northern Arizona University
Box 15066
Flagstaff, AZ 86011
(602) 523-7405

Mr, Dan Anderson
Arizona Department of Economic
Security
Research Administration
P.O. Box 6123
Phoenix, AZ 85005
(602) 255-3616

Mr. Mobin Qaheri
Senior Economic Specialist
Planning and Policy Development
AZ Department of Commerce
Suite 1400, 3800 North Central
Phoenix, AZ 85012
(602) 280-1321

Mr. Henry C. Reardon
Chief Economist
Joint Legislative Budget
Committee
1716 West Adams Street
Phoenix, AZ 85007
(602) 542-5491

Mr. Tom R. Rex
Manager of Research Support
Arizona State University
Center for Business Research
Tempe, AZ 85287
(602) 965-3961

Ms. Pia Montoya
Division of Economic and
Business Research
University of Arizona
B.P.A. Building
Tucson, AZ 85721
(602) 621-2155

ARKANSAS

Dr. Donald Market
Director
Bureau of Business and Economic
Research
College of Business Administration
Room 443
University of Arkansas
Fayetteville, AR 72201
(501) 575-4151

Mrs Alma Holbrook
Employment Security Division
State Capital Mall
P.O. Box 2981
Little Rock, AR 72203
(501) 682-3198

Ms. Neva Wayman
UALR Library 512
Regional Economic Analysis
2801 South University
Little Rock, AR 72204
(501) 569-8551

CALIFORNIA

Mr. Fred Gey
U.C. Data Archive and Technical
Assistance
University of California - Berkeley
2538 Channing Way
Berkeley, CA 94720
(415) 642-1472

Dr. John G. Sanzone
Director University Center for
Economic Development and
Planning
California State University at
Chico
Chico, CA 95929
(916) 898-4598

Ms. Patricia C. Inouye
Shields Library
Government Documents Department
University of California at Davis
Davis, CA 95616
(916) 752-1624

Mrs Pauline P. Sweezey
Financial and Economic Research
Department of Finance
915 L Street, 8th Floor
Sacramento, CA 95814
(916) 322-2263

COLORADO

Ms. Gin Hayden
System Analyst
Graduate School of Business
Administration
Campus Box 420
University of Colorado
Boulder, CO 80309
(303) 492-8227

Mr. Ken Anderson
Labor Market Information
Division of Employment and
Training
251 East 12th Avenue
Denver, CO 80203
(303) 866-6328

Dr. Reid T. Reynolds
Colorado Division of Local
Government
1313 Sherman Street
Room 520
Denver, CO 80203
(303) 866-2156

Mr. Curt Wiedeman
Office of State Planning and
Budget
111 State Capital Building
Denver, CO 80203
(303) 866-3319

Dr. John W. Green
Department of Economics
University of Northern Colorado
Greeley, CO 80639
(303) 351-2739

CONNECTICUT

Mr. Ming J. Wu
Budget Specialist
Office of Policy and Management
80 Washington Street
Hartford, CT 06106
(203) 566-8342

Mr. Richard L. Vannuccini
Connecticut Labor Department
Employment Security Division
200 Folly Brook Boulevard
Wethersfield, CT 06109
(203) 566-2120

DELAWARE

Mr. Doug Clendaniel
Delaware Development Office
P.O. Box 1401
99 King Highway
Dover, DE 19903
(302) 736-4271

Mr. Robert Schulz
Delaware Department of Labor
P.O. Box 9029
Newark, DE 19714
(302) 368-6960

Mr. Edgar Williamson
Senior Assistant Librarian
University of Delaware Library
Newark, DE 19717
(302) 451-2432

Mr. James Craig
Department of Finance
Bureau of Economics and Statistics
Carvel State Office Building
820 North French Street
Wilmington, DE 19801
(302) 571-3324

DISTRICT OF COLUMBIA

Mr. Gan Ahuja
D.C. Planning Office
Presidential Building
Suite 570
415 12th Street, N.W.
Washington, DC 20004
(202) 727-6533

Ms. Lori Hunter
Department of Finance and Revenue
Municipal Center, Room 4130
300 Indiana Avenue, N.W.
Washington, DC 20001
(202) 727-6027

Mr. Paul Des Jardin
Metropolitan Washington Council
of Governments, Suite 300
777 North Capital Street, N.E.
Washington, DC 20002
(202) 223-6800

FLORIDA

Mr. David Heisser
Government Publications
University of Miami Library
P. O. Box 248214
Coral Gables, FL 33124
(305) 284-3155

Ms. Janet Galvez
Bureau of Economic and Business
Research
University of Florida
221 Matherly Hall
Gainesville, FL 32611
(904) 392-0171

Ms. Rebecca Rust
Chief
Bureau of Labor Market
Information
200 Hartman Building
2012 Capital Circle S.E.
Tallahassee, FL 32399
(904) 488-1048

Ms. Sarah Voyles
Executive Office of the Governor
The Capitol
Tallahassee, FL 32399
(904) 487-2814

Librarian
Bureau of Economic Analysis
Florida Department of Commerce
Research Library
Collins Building
Tallahassee, FL 32399
(904) 487-2971

Mr. Thomas Charles
Research Associate
Center for Economic and
Management Research
COBA BSN 3403
University of South Florida
4202 E. Fowler Avenue
Tampa, FL 33620
(813) 974-4266

GEORGIA

Dr. Albert W. Niemi, Jr.
Selig Center for Economic Growth
Brooks Hall
University of Georgia
Athens, GA 30602
(404) 542-4085

Mrs. Louise Allen
Assistant Director
Research Division
Georgia Department of Industry and
Trade
285 Peachtree Center Avenue
Atlanta, GA 30303
(404) 656-7655

Ms. Robin Kirkpatrick
Georgia Office of Planning
and Budget
Room 640
254 Washington Street S.W.
Atlanta, GA 30334
(404) 656-0911

Mr. Richard Leacy
Head
Government Documents and Maps
Department
Georgia Institute of Technology
Library
Atlanta, GA 30332
(404) 894-4519

Mr. Milton Martin
Director
Georgia Department of Labor
Labor Information Systems
254 Washington Street S.W.
Atlanta, GA 30334
(404) 656-3177

Dr. Donald Ratajczak
Economic Forecast
Georgia State University
30 Pryor, Room 715
Atlanta, GA 30303
(404) 651-3282

Mr. Tshai Alemayehu
Bureau of Business Research
School of Business
Savannah State College
Savannah, GA 31404
(912) 356-2830

HAWAII

Dr. Richard Y. P. Joun
Department of Business and
Economic Development
P.O. Box 2359
Honolulu, HI 96804
(808) 586-2470

Mr. Robert Koike
Hawaii Department of Taxation
P.O. Box 259
Honolulu, HI 96809
(808) 548-7635

Mr. Fred Pang
Department of Labor and Industrial
Relations
830 Punchbowl Street
Honolulu, HI 96813
(808) 548-7639

IDAHO

Mr. James L. Adams
Chief
Research and Analysis
Idaho Department of Employment
317 Main Street
Boise, ID 83735
(208) 334-6169

Mr. Michael Ferguson
Division of Financial Management
State of Idaho
State House, Room 122
Boise, ID 83720
(208) 334-2950

Mr. Alan Porter
Idaho Department of Commerce
700 West State Street
Boise, ID 83720
(208) 334-2470

Mr. Charles L. Skoro
Economics Department
Boise State University
1910 University Drive
Boise, ID 83725
(208) 385-1117

Mr. Lawrence H. Merk
Director
Center for Business Development
and Research
College of Business and Economics
University of Idaho
Moscow, ID 82843
(208) 885-6611

Mr. Paul Zelus
Center for Business Research
and Services
Idaho State University
Campus Box 8450
Pocatello, ID 83725
(208) 236-3050

ILLINOIS

Dr. Roger Beck
Chairman
Southern Illinois University
Department of Agribusiness
Economics
Carbondale, IL 62901
(618) 453-2421

Ms. Martha Greene
Bureau of Economic and Business
Research
University of Illinois
1206 South Sixth Street
Champaign, IL 61820
(217) 244-3099

Mr. Wallace Biermann
Division of Research and Analysis
Department of Commerce and
Community Affairs
620 East Adams
Springfield, IL 62701
(217) 782-1438

Ms. Sue Ebetsch
Coordinator
Illinois State Data Center
Cooperative
Illinois Bureau of the Budget
605 Stratton Building
Springfield, IL 62706
(217) 782-3500

INDIANA

Professor Morton J. Marcus
Research Economist
Division of Research
School of Business
Indiana University
Bloomington, IN 47405
(812) 855-5507

Ms. Terry Creeth
Indiana Business Research Center
801 W. Michigan Street BS4015
Indianapolis, IN 46202
(317) 274-2204

Mr. Richard Dierdorf
Indiana Department of Employment
and Training
Labor Market Analysis
10 North Senate Avenue
Indianapolis, IN 46204
(317) 232-8536

Ms. Roberta Eads
Indiana State Data Center
Indiana State Library
140 North Senate Avenue
Indianapolis, IN 46204
(317) 232-3735

Mr. Robert Lain
Research Office
Indiana Department of Commerce
One N.Capital, Suite 700
Indianapolis, IN 46204
(317) 233-3559

IOWA

Dr. Daniel Otto
Extension Economist
Iowa State University
560 Heady Hall
Ames, IA 50111
(515) 294-7518

Mr. Ron Amosson
Office of the State Comptroller
Iowa State Capital, Room 12
Des Moines, IA 50319
(515) 281-3078

Ms. Judy K. Erickson
Iowa Department of Employment
Services
Audit and Analysis Department
1000 East Grand Avenue
Des Moines, IA 50309
(515) 281-3439

Mr. Harvey Siegelman
State Economist
Department of Economic
Development
200 East Grand Avenue
Des Moines, IA 50309
(515) 242-4868

Professor Charles H. Whiteman
Department of Economics
University of Iowa
Iowa City, IA 52242
(319) 335-0835

KANSAS

Ms. Thelma Helyar
Institute for Public Policy and
Business Research
The University of Kansas
607 Blake Hall
Lawrence, KS 66045
(913) 864-3701

Mr. Marc Galbraith
Kansas State Library
State Capitol Building
Room 343-N
Topeka, KS 66612
(913) 296-3296

Mr. William H. Layes
Chief
Kansas Department of Human
Resources
Labor Market Information Services
401 SW Topeka Boulevard
Topeka, KS 66603
(913) 296-5058

Ms. Debbie McCaskill
Director
Market Analysis
Kansas Department of Commerce
Suite 500, 400 West 8th
Topeka, KS 66603
(913) 296-3760

Ms. Carelene Hill Forrest
Director
Center for Economic Development
and Business Research
School of Business
Campus Box 48
Wichita State University
1845 Fairmont
Wichita, KS 67208
(316) 689-3225

KENTUCKY

Mr. Ed Blackwell
LMI Supervisor
Kentucky Department of
Employment Services
Cabinet of Human Resources
2nd Floor East
275 East Main Street
Frankfort, KY 40621
(502) 564-7976

Mr. James A. Street
Deputy Director
Office of Financial Management
and Economic Analysis
Finance and Administration
Cabinet
318 Capitol Annex
Frankfort, KY 40601
(502) 564-2924

Mr. Roy Sigafus
Center for Business and Economic
Research
University of Kentucky
301 Mathews Building
Lexington, KY 40506
(606) 257-7678

Mr. Ron Crouch
Director
Urban Studies Center
University of Louisville
Louisville, KY 40292
(502) 588-6626

LOUISIANA

Ms. Karen Paterson
Louisiana State Planning Office
P.O. Box 94095
Baton Rouge, LA 70804
(504) 342-7410

Mr. Oliver Robinson, III
Director
Louisiana State Department of
Employment and Training
P.O. Box 94094
Capitol Station
1001 North 23rd Street
Baton Rouge, LA 70804
(504) 342-3143

Dr. Loren Scott
College of Business Administration
Louisiana State University
Baton Rouge, LA 70803
(504) 388-3779

Dr. Jerry L. Wall
Director
Center for Business and Economic
Research
Northeast Louisiana University
Monroe, LA 71209
(318) 342-1215

Mr. Vincent Maruggi
Division of Business and Economic Research
University of New Orleans
New Orleans, LA 70148
(504) 286-6248

Dr. James Robert Michael
Director
Research Division
Louisiana Tech University
P.O. Box 10318
Ruston, LA 71272
(318) 257-3701

MAINE

Mr. Stephen J. Adams
State Economist
Maine State Planning Office
184 State Street
State House Station 38
Augusta, ME 04333
(207) 289-3261

Mr. Ray A. Fongemie
Director
Division of Economic Analysis and Research
Maine Department of Labor
20 Union Street
Augusta, ME 04330
(207) 289-2271

Mr. Philip Gardner
Research Division
Bureau of Taxation
Station 24
Augusta, ME 04333
(207) 289-4702

Mr. James H. Breece
New England Electronic Data Center
Steven Hall
Department of Economics
University of Maine at Orono
Orono, ME 04469
(207) 581-1862

Mr. Robert C. McMahon
Center for Business and Economic Research
University of Southern Maine
96 Falmouth Street
Portland, ME 04103
(207) 780-4308

MARYLAND

Mr. Pat Arnold
Director
Labor Market Analysis / Information
Maryland Department of Economic and Employment Development
1100 North Eutaw Street
Baltimore, MD 21201
(301) 333-5000

Mr. Michael Lettre
Assistent Director
Maryland Office of Planning
301 West Preston Street
Baltimore, MD 21201
(301) 225-4450

Dr. Lester Salamon
Institute for Policy Studies
Johns Hopkins University
Shriver Hall
Baltimore, MD 21218
(301) 338-7174

Ms. Peggy Dalton
Guild Center
Economics Department
Frostburg State University
Frostburg, MD 21532
(301) 689-4386

MASSACHUSETTS

Mr. Sephen P. Coelen
Director
Massachusetts Institute for Social
and Economic Research
128 Thompson Hall
University of Massachusetts
Amherst, MA 01003
(413) 545-3460

Ms. Carla Miller
Massachusetts State Data Center
Room 50, State House
c/o Miser: P.O. Box 219
Boston, MA 02133
(617) 727-4537

Mr. Gregory Perkins
Research Directorate
Boston Redevelopment Authority
1 City Hall Square
Boston, MA 02201
(617) 242-7400

Mr. Elliot Winer
Massachusetts Department of
Employment and Training
Economic Research and Analysis
Department
C. F. Hurley Building
2nd Floor
Boston, MA 02114
(617) 727-7428

MICHIGAN

Mr. Von D. Logan
Director
Bureau of Research and Statistics
Michigan Employment Security
Commission
7310 Woodward Avenue
Detroit, MI 48202
(313) 876-5427

Ms. Bettie Landauer-Menchik
Michigan State University
Center for the Redevelopment of
Industrialized States
403 Olds Hall
East Lansing, MI 48824
(517) 353-3255

Mr. Warren G. Gregory
House Fiscal Agency
Michigan House of Representatives
Manufacturers Bank Building
P.O. Box 30014
Lansing, MI 48909
(517) 373-8080

Mr. Mark Haas
Michigan Department of Commerce
P.O. Box 30225
525 West Ottawa
Lansing, MI 48909
(517) 373-4600

Mr. Dan Kitchel
Michigan Department of Treasury,
Office of Revenue and Tax
Analysis
Treasury Building
Lansing, MI 48922
(517) 373-2958

Bureau of Economic Analysis State User Groups   *XI. Business Data Directory*

MINNESOTA

Dr. Jerrold Peterson
Acting Director
Bureau of Business and Economic
Research
University of Minnesota at Duluth
Room 150 SBE
Duluth, MN 55812
(218) 726-7298

Mrs. Wendy Treadwell
Machine Readable Data Center
University of Minnesota
Room 1 Wilson Library
309 19th Avenue South
Minneapolis, MN 55455
(612) 624-4389

Mr. Ned Chottepanda
Minnesota Department of Jobs and
Training
Room 517
390 N. Robert Street
St. Paul, MN 55101
(612) 296-6545

Mr. Adam G. Marsnik
Legislative Reference Library
645 State Office Building
100 Constitution Avenue
St. Paul, MN 55101
(612) 296-0586

Mr. David Rademacher
Minnesota State Planning Agency
Office of State Demographer
300 Centennial Office Building
St. Paul, MN 55155
(612) 296-3255

Mr. Carroll Rock
State Agricultural Statistician
Agricultural Statistics Division
Minnesota Department of
Agriculture
Box 7068
St. Paul, MN 55107
(612) 296-2230

MISSISSIPPI

Dr. R. Eric Reidenbach
Center for Business Development
and Research
University of Southern Mississippi
Southern Station, Box 5094
Hattiesburg, MS 39406
(601) 266-7011

Mr. Raiford G. Crews
Chief
Labor Market Information
Mississippi Employment Security
Commission
1520 West Capitol Street
P.O. Box 1699
Jackson, MS 39215
(601) 961-7424

Ms. Missy Lee
Information Services Library
Mississippi Institutions of Higher
Learning
3825 Ridgewood Road
Jackson, MS 39211
(601) 982-6314

Dr. J. William Rush
Associate Dean
Office of External Affairs
Mississippi State University
P.O. Drawer 5288
Mississippi State, MS 39762
(601) 325-3817

Dr. Max W. Williams
Center for Population Studies
University of Mississippi
Bondurant Building, Room 3W
University, MS 38677
(601) 232-7288

MISSOURI

Dr. Edward H. Robb
Director
B & PA Research Center
University of Missouri at Columbia
10 Professional Building
Columbia, MO 65211
(314) 882-4805

Ms. Kate Graf
Census Data Center
Missouri State Library
P.O. Box 387
Jefferson City, MO 65102
(314) 751-2361

Mr. Thomas Kruckemeyer
Planner
Division of Budget and Planning
Office of Administration
Capitol Building, Room 124
Jefferson City, MO 65102
(314) 751-2345

MONTANA

Ms. Patricia Roberts
Census and Economic Information
Center
Montana Department of Commerce
1424 Ninth Avenue
Helena, MT 59620
(406) 444-4393

Mr. Paul E. Polzin
Director
Bureau of Business and Economic
Research
University of Montana
Missoula, MT 59812
(406) 243-5113

NEBRASKA

Mr. Gary B. Heinicke
Administrator
Research Division
Nebraska Department of Revenue
Box 94818
Lincoln, NE 68509
(402) 471-2971

Dr. F. Charles Lamphear
Bureau of Business Research
University of Nebraska - Lincoln
College of Business Administration
Lincoln, NE 68588
(402) 472-2334

Mr. Tim K. Himberger
Center for Public Affairs Research
University of Nebraska at Omaha
Omaha, NE 68182
(402) 595-2311

NEVADA

Mr. James S. Hanna
Chief
Employment Security Research
500 East Third Street
Carson City, NV 89713
(702) 885-4550

Ms. Betty McNeal
Nevada State Library and Archives
Capitol Complex
Carson City, NV 89710
(702) 885-5160

Mr. Martin Kyte
Operations Manager
Bureau of Business and Economic Research
University of Nevada at Reno
Reno, NV 89557
702) 784-4820

NEW HAMPSHIRE

Mr. George Bruno
New Hampshire Department of Resources and Economic Development
105 Loudon Road
Concord NH 03302
(603) 271-2591

Mr. Thomas J. Duffy
Senior Planner
Office of State Planning
2 1/2 Beacon Street
Concord, NH 03301
(603) 271-2155

Mr. William E. Pillsbury, Jr.
New Hampshire Office of Industrial Development
P.O. Box 856
Concord, NH 03301
(603) 271-2591

NEW JERSEY

Librarian
Bureau of Economic Research
Rutgers State University of New Jersey, New Jersey Hall
New Brunswick, NJ 08903
(201) 932-8019

Ms. Connie O. Hughes
Assistant Director
Division of Labor Market and Demographic Research
NJ Department of Labor
CN 388 John Fitch Plaza
Trenton, NJ 08625
(609) 984-2593

Mr. Henry A. Watson
Director
Center for Health Statistics
Room 405
New Jersey Department of Health
CNN 360
Trenton, NJ 08625
(609) 984-6703

NEW MEXICO

Mr. Gerard P. Bradley
Economic Analyst
Economic Research and Analysis
Employment Security Division
NM Department of Labor
Box 1928, 401 Broadway, NE
Albuquerque, NM 87103
(505) 841-8645

Mr. Brian McDonald
Director
Bureau of Business and Economic
Research
University of New Mexico
1920 Lomas N.E.
Albuquerque, NM 87131
(505) 277-2216

Dr. James T. Peach
Department of Economics
New Mexico State University
Box 30001 / Department 3CQ
Las Cruces, NM 88003
(505) 646-3113

Ms. Carol Selleck
Economic Development and
Tourism Department
Joseph Montoya Building
Sante Fe, NM 87503
(505) 827-0276

NEW YORK

Mr. William T. Grainger
Bureau of Economic and
Demographic Information
New York State Department of
Economic Development
One Commerce Plaza, RM 910
Albany, NY 12245
(518) 474-1161

Mr. Barclay G. Jones
Director
Cornell University
CISER/PURS
106 West Sibley Hall
Ithaca, NY 14853
(607) 256-4331

Mr. Ronald J. Ortiz
Demographer
Population Division
New York City Dept. of Planning
22 Reade Street - 4 West
New York, NY 10007
(212) 720-3446

NORTH CAROLINA

Dr. Rick Kirkpatrick
Director
Bureau of Business and Economic
Research, John A. Walker College
of Business
Appalachian State University
Boone, NC 2868
(704 262-6127

Dr. James F. Smith
Graduate School of Business
Administration
University of North Carolina
Chapel Hill, NC 27599
(919) 962-3176

Dr. John E. Connaughton
Director
Center for Business and Economic
Research, University of North
Carolina - Charlotte
Room 232, Friday Building
Highway 49
Charlotte, NC 28233
(704) 547-2185

Mr. Delos Monteith
Center for Improving
Mountain Living
Cullowhee, NC 28723
(704 227-7492

Director
Tax Research Division
North Carolina
Department of Revenue
P. O. Box 25000
Raleigh, NC 27640
(919) 733-4549

Mr. Gregory B. Sampson
Director
Bureau of Employment Security
Commission of North Carolina
P.O. Box 24903
Raleigh, NC 27611
(919) 733-2936

Ms. Francine J. Stephenson
Manager
State Data Center
Office of State Planning
116 West Jones Street
Raleigh, NC 27603
(919) 733-4131

NORTH DAKOTA

Mr. Sid Bender
Office of State Tax Commissioner
State Capitol
Bismarck, ND 58505
(701) 224-3402

Mr. Tom Pederson
Director
Research and Statistics
Job Service North Dakota
P.O. Box 1537
Bismarck, ND 58502
(701) 224-2868

Ms. Carol Vavrosky
Department of Agricultural
Economics
North Dakota State University
Fargo, ND 58105
(701) 237-7400

Dr. Scot A. Stradley
Director
Bureau of Business and Economic
Research
University of North Dakota
P.O. Box 8255
University Station
Grand Forks, ND 58202
(701) 777-2637

OHIO

Dr. Steven Howe
SW Ohio Regional Data Center
University of Cincinnati
Mail Location #132
Cincinnati, OH 45221
(513) 556-5028

Dr. Leroy Hushak
Department of Agricultural
Economics, and Rural Sociology
Ohio State University
2120 Fyffe Road
Columbus, OH 43210
(614) 292-3548

Mr. Larry Less
Projections Coordinator
Labor Market Information Division
Ohio Bureau of Employment
Services
1160 Dublin Road
Columbus, OH 43215
(614) 644-2689

Ms. Geraldine Waller
Ohio Data Users Center
Box 1001, 26th Floor
Columbus, OH 43266
(614) 466-2115

Mr. Paul J. Kozlowski
Toledo Economic Information
System
University of Toledo
Stranahan Hall
Toledo, OH 43606
(419) 537-2067

OKLAHOMA

Mr. John McCraw
Center for Economic and
Management Research
University of Oklahoma
Room 4, 307 W. Brooks Street
Norman, OK 73019
(405) 325-2931

Mr. Roger Jacks
Office of Economic Analysis
Oklahoma Security Commission
305 Will Rogers Memorial Office
Building
Oklahoma City, OK 73105
(405) 557-7106

Mr. Jeff Wallace
Oklahoma State Data Center
Oklahoma Department of Commerce
P.O. Box 26980
Oklahoma City, OK 73126
(405) 843-9770

Dr. Ahmed Abo-Basha
Office of Business and Economic
Research
Oklahoma State University
345 College of Business
Administration
Stillwater, OK 74078
(405) 744-5125

OREGON

Mr. Stanley D. Miles
Extension Economist
AREC Extension Economic
Information Office / 219
Ballard Extension Hall
Oregon State University
Corvallis, OR 97331
(503) 754-2942

Ms. Karen Seidel
Bureau of Governmental Research
and Service
University of Oregon
P.O. Box 3177
Eugene, OR 97403
(503) 686-5232

Mr. Arthyr Ayre
Oregon Economic Development
Department
775 Summer Street, N.E.
Salem, OR 97310
(503) 373-1200

Ms. Virlena Crosley
Assistant Administrator
Employment Division
Department of Human Resources
875 Union Street, N.E.
Salem, OR 97311
(503) 378-3220

Mr. Paul Warner
State Economist
Executive Department
155 Cottage Street, N.E.
Salem, OR 97310
(503) 378-3405

PENNSYLVANIA

Mr. Carl Thomas
Director
Bureau of Research and Statistics
Pennsylvania Department of Labor
and Industry
7th and Forster Streets
Harrisburg, PA 17121
(717) 787-3265

Mr. Michael T. Behney
Pennsylvania State Data Center
Pennsylvania State University
Middletown, PA 17057
(717) 948-6336

Dr. Robert Sechrist
Director
Institute for Research and
Community Service
Indiana University of Pennsylvania
110 Stright Hall
Indiana, PA 15705
(412) 357-2251

R. L. Bangs
Research Associate
University of Pittsburgh
121 University Place
Pittsburgh, PA 15260
(412) 624-3856

Dr. Rodney A. Erickson
Director
Center for Regional Business
Analysis
Pennsylvania State University
108 Business Administration
Building II
University Park, PA 16802
(814) 865-7669

RHODE ISLAND

Mr. Vincent K. Harrington
Rhode Island Department of
Economic Development
7 Jackson Walkway
Providence, RI 02903
(401) 277-2601

Mr. Robert J. Langlais
Administrator
Labor Market Information and
Management Services
Rhode Island Department
of Employment and Training
101 Friendship Street
Providence, RI 02903
(401) 277-3704

SOUTH CAROLINA

Ms. Betsy Jane Clary
Director
Bureau of Economic and Business
Research
School of Business and Economics
College of Charleston
9 Liberty Street
Charleston, SC 29424
(803) 792-8107

Dr. William Gillespie
Division of Research and Statistics
Room 440
Rembert Dennis Building
1000 Assembly Street
Columbia, SC 29201
(803) 734-3798

Dr. Randy Martin
Director
Division of Research
College of Business Administration
University of South Carolina
Columbia, SC 29208
(803) 777-2510

South Carolina Employment
Security Commission
Labor Market Information Division
1550 Gadsden Street
P.O. Box 995
Columbia, SC 29202
(803) 737-2660

SOUTH DAKOTA

Ms. Mary Susan Vickers
Chief
Labor Market Information Center
South Dakota
Department of Labor
Box 4730, 420 South Roosevelt
Aberdeen, SD 57402
(605) 622-2314

Mr. Donald Lewis
Business Research Bureau
School of Business
University of South Dakota
414 East Clark
Vermillion, SD 57069
(605) 677-5287

Bureau of Economic Analysis State User Groups   *XI. Business Data Directory*

TENNESSEE

Dr. David Hake
Director
Center for Business and Economic
Research
University of Tennessee
Knoxville, TN 37996
(615) 974-5441

Mr. Ed Jernigan
State Data Center
Bureau of Business and Economic
Research
Memphis State University
Memphis, TN 38152
(901) 454-2281

Mr. Bobby N. Corcoran
Middle Tennessee State University
P.O. Box 485
Murfreesboro, TN 37132
(615) 898-2520

Mr. Charles Brown
Planning Analyst
State Planning Office
309 John Sevier Building
500 Charlotte Avenue
Nashville, TN 37243
(615) 741-1676

Mr. Joe S. Cummings
Tennessee Department of
Employment Security
500 James Robertson Parkway
Eleventh Floor
Nashville, TN 37245
(615) 741-2284

TEXAS

Mr. F.G. Bloodworth
Chief
Water Uses and Projections Section
Texas Water Development Board
P.O. Box 13231
Capitol Station
Austin, TX 78711
(512) 463-7940

Mr. Mark Hughes
Texas Employment Commission
TEC Building
Austin, TX 78778
(512) 463-2326

Ms. Susan Tully
Texas State Data Center
Texas Department of Commerce
P.O. Box 12728
Capitol Station
Austin, TX 78711
(512) 472-5059

Mrs. Rita Wright
Bureau of Business Research
University of Texas at Austin
P.O. Box 7459
University Station
Austin, TX 78713
(512) 471-5180

Mr. Jesse Acosta
Supervisor
Department of Planning
Two Civic Center Plaza
El Paso, TX 79901
(915) 541-4721

147

Dr. Robert F. Hodgin
University of Houston
Clear Lake City
UH - Clear Lake Box 288
2700 Bay Area Boulevard
Houston, TX 77058
(713) 488-7170

Ms. Michelle Kretzschmar
Institute for Studies in Business
College of Business
University of Texas at
San Antonio
San Antonio, TX 78249
(512) 691-4317

UTAH

Mr. Kenneth E. Jensen
Labor Economist
Utah Department of Employment
Security
P.O. Box 11249
Salt Lake City, UT 84147
(801) 533-2372

Dr. R. Thayne Robson
Director
Bureau of Economic and Business
Research
University of Utah
KDGB 401
Salt Lake City, UT 84112
(801) 581-7274

Ms. Linda Smith
State Data Center
116 State Capitol Building
Salt Lake City, UT 84114
(801) 581-7274

Mr. Thomas M. Williams
Senior Economist
Utah State Tax Commission
519 Heber M.Wells Building
160 East 300 South
Salt Lake City, UT 84134
(801) 530-6093

VERMONT

Mr. Jeff Carr
Office of Policy Research
Coordination
Pavilion Office Building
109 State Street
Montpelier, VT 05602
(802) 828-3326

Mr. Michael Griffin
Chief
Office of Labor Market Information
Vermont Department of
Employment and Training
P.O. Box 488
Montpelier, VT 05601
(802) 828-4323

VIRGINIA

Mr. Randy Austin
Department of Agricultural
Economics
Virginia Polytechnic Institute and
State University
Blacksburg, VA 24061
(703) 231-7936

Dr. John L. Knapp
Research Director
Business and Economic Section
Center for Public Service
University of Virginia
Fourth Floor, Dynamics Building
2015 Ivy Road
Charlottesville, VA 22903
(804) 924-3434

Dr. Roger R. Stough
George Mason University Institute
of Public Policy
Krug Hall 205
Fairfax, VA 22030
(703) 993-2280

Dr. Richard A Phillips
Editor
Hampton Roads Economic Report
School of Business Administration
Old Dominion University
Norfolk, VA 23508
(804) 440-4713

Mr. Thomas McGraw
Senior Economist
Economic Information Services
Virginia Employment Commission
P.O. Box 1358
Richmond, VA 23211
(804) 786-3177

Mr. Larry Robinson
Department of Planning and Budget
445 9th Street Office Building
P.O. Box 1422
Richmond, VA 23211
(804) 786-8624

Mr. Roy L. Pearson
Director
Bureau of Business Research
School of Business Administration
College of William and Mary
Williamsburg, VA 23185
(804) 221-2930

WASHINGTON

Mr. Byron Angel
Forecasting Division
Office of Financial Management
Insurance Building Room 450
AQ-44
Olympia, WA 98504
(206) 586-2478

Mr. Bret Bertolin
Office of the Forecast Council
Mail Stop FJ-33
Olympia, WA 98504
(206) 586-6736

Mr. Gary Smith
Extension Economist
Washington State University
203C Hulbert Hall
Pullman, WA 99164
(509) 335-2852

Professor Philip J. Bourque
Professor of Business Economics
Graduate School of Business
Administration
University of Washington
Seattle, WA 98195
(206) 543-4484

WEST VIRGINIA

Mr. Fred Cutlip
Director
Community Development Division
Building 6, Room B-553
Capitol Complex
Charleston, WV 25305
(304) 348-4010

Mr. Edward F. Merrifield
Assistant Director
West Virginia Division of
Employment Security
Labor and Economic Research
112 California Avenue
Charleston, WV 25305
(304) 348-2660

Mr. Allan L. Mierke
Assistant Tax Commissioner
Department of Tax and Revenue
P.O. Box 2389
Charleston, WV 25328
(304) 348-3478

Ms. Linda Culp
Bureau of Business Research
West Virginia University
323 Business and Economic
Building
P.O. Box 6025
Morgantown, WV 26506
(304) 293-7832

WISCONSIN

Ms. Sandra Koch
Director
Bureau of Business and
Economic Research
University of Wisconsin -
La Crosse
204 North Hall
1725 State Street
La Crosse, WI 54601
(608) 785-8500

Mr. Hartley J. Jackson
Director
Department of Industry
Labor and Human Relations
201 E. Washington Avenue
P.O. Box 7944
Madison, WI 53707
(608) 266-7034

Mr. Robert Naylor
Demographic Services Center
Department of Administration
101 S. Webster Street, 6th Floor
P.O. Box 7868
Madison, WI 53707
(608) 266-1927

Mr. Gene Schubert
Wisconsin Department of Revenue
125 South Webster Street
Madison, WI 53702
(608) 266-8132

Dr. William A. Strang
Associate Dean
Office of Business Research and
Services, School of Business
University of Wisconsin - Madison
1155 Observatory Drive
Madison, WI 53706
(608) 262-1550

WYOMING

Mr. Thomas N. Gallagher
Research and Planning Division
Department of Employment
100 West Midwest
P.O. Box 2760
Casper, WY 82602
(307) 235-3646

Ms. Mary Byrnes
Wyoming Department of
Administration and Information
Economic Analysis Division
Room 327E, Emerson Building
Cheyenne, WY 82002
(307) 777-7504

Mr. G. Fred Doll
Director
Survey Research Center
University of Wyoming
P.O. Box 3925
Laramie, WY 82071
(307) 766-5141

Dr. Clynn Phillips
Associate Director
Dept. of Agricultual Economics
University of Wyoming
P.O. Box 3354
University Station
Laramie, WY 82071
(307) 766-2178

The Bureau of Labor Statistics: Regional Offices

Regional offices of The Bureau of Labor Statistics (BLS) supply information through a publication program, which includes press releases, periodicals, reports, and bulletins. Some BLS material is available on magnetic tapes, diskettes, and microfiche. In addition to providing data covering the broad field of labor economics, BLS also provides a number of services for users of their data. These include, the release of certain categories of unpublished data; the development of special surveys and tabulations; the duplication of machine-readable data files; and the sale of statistical software programs. BLS regional offices also provide consultant services on survey techniques and the production, use and limitations of BLS data.

The Bureau of Labor Statistics Regional Offices

ALABAMA
Bureau of Labor Statistics
1371 Peachtree Street N.E.
Atlanta, GA 30367
(404) 347-4416

ALASKA
Bureau of Labor Statistics
71 Stevenson Street
P.O. Box 193766
San Francisco, CA 94119
(415) 744-6600

AMERICAN SOMOA
Bureau of Labor Statistics
71 Stevenson Street
P.O. Box 3766
San Francisco, CA 94119
(415) 744-6600

ARIZONA
Bureau of Labor Statistics
71 Stevenson Street
P.O. Box 3766
San Francisco, CA 94119
(415) 744-6600

ARKANSAS
Bureau of Labor Statistics
Federal Building, Room 221
525 Griffin Street
Dallas, TX 75202
(214) 767-6970

CALIFORNIA
Bureau of Labor Statistics
71 Stevenson Street
P.O. Box 3766
San Francisco, CA 94119
(415) 744-6600

COLORADO
Bureau of Labor Statistics
911 Walnut Street
Kansas City, MO 64106
(816) 426-2481

CONNECTICUT
Bureau of Labor Statistics
1 Congress Street
10th Floor
Boston, MA 02114
(617) 565-2327

DELAWARE
Bureau of Labor Statistics
3535 Market Street
P.O. Box 13309
Philadelphia, PA 19101
(215) 596-1154

DISTRICT OF COLUMBIA
Bureau of Labor Statistics
3535 Market Street
P.O. Box 13309
Philadelphia, PA 19101
(215) 596-1154

FLORIDA
Bureau of Labor Statistics
1371 Peachtree Street N.E.
Atlanta, GA 30367
(404) 347-4416

GUAM
Bureau of Labor Statistics
71 Stevenson Street
P.O. Box 3766
San Francisco, CA 94119
(415) 744-6600

GEORGIA
Bureau of Labor Statistics
1371 Peachtree Street N.E.
Atlanta, GA 30367
(404) 347-4416

HAWAII
Bureau of Labor Statistics
71 Stevenson Street
P.O. Box 3766
San Francisco, CA 94119
(415) 744-6600

IDAHO
Bureau of Labor Statistics
71 Stevenson Street
P.O. Box 3766
San Francisco, CA 94119
(415) 744-6600

ILLINOIS
Bureau of Labor Statistics
9th Floor
Federal Office Building
230 S. Dearborn Street
Chicago, IL 60604
(312) 353-1880

INDIANA
Bureau of Labor Statistics
9th Floor
Federal Office Building
230 S. Dearborn Street
Chicago, IL 60604
(312) 353-1880

IOWA
Bureau of Labor Statistics
911 Walnut Street
Kansas City, MO 64106
(816) 426-2481

KANSAS
Bureau of Labor Statistics
911 Walnut Street
Kansas City, MO 64106
(816) 426-2481

KENTUCKY
Bureau of Labor Statistics
1371 Peachtree Street N.E.
Atlanta, GA 30367
(404) 347-4416

LOUISIANA
Bureau of Labor Statistics
Room 221, Federal Building
525 Griffin Street
Dallas, TX 75202
(214) 767-6970

MISSOURI
Bureau of Labor Statistics
911 Walnut Street
Kansas City, MO 64106
(816) 426-2481

MONTANA
Bureau of Labor Statistics
911 Walnut Street
Kansas City, MO 64106
(816) 426-2481

NEBRASKA
Bureau of Labor Statistics
911 Walnut Street
Kansas City, MO 64106
(816) 426-2481

NEVADA
Bureau of Labor Statistics
71 Stevenson Street
P.O. Box 3766
San Francisco, CA 94119
(415) 744-6600

NEW HAMPSHIRE
Bureau of Labor Statistics
1 Congress Street
10th Floor
Boston, MA 02114
(617) 565-2327

NEW JERSEY
Bureau of Labor Statistics
Room 808
201 Varick Street
New York, NY 10014
(212) 337-2400

NEW MEXICO
Bureau of Labor Statistics
Room 221, Federal Building
525 Griffin Street
Dallas, TX 75202
(214) 767-6970

NEW YORK
Bureau of Labor Statistics
Room 808
201 Varick Street
New York, NY 10014
(212) 337-2400

NORTH CAROLINA
Bureau of Labor Statistics
1371 Peachtree Street N.E.
Atlanta, GA 30367
(404) 347-4416

NORTH DAKOTA
Bureau of Labor Statistics
911 Walnut Street
Kansas City, MO 64106
(816) 426-2481

OHIO
Bureau of Labor Statistics
9th Floor
Federal Office Building
230 S. Dearborn Street
Chicago, IL 60604
(312) 353-1880

OKLAHOMA
Bureau of Labor Statistics
Room 221, Federal Building
525 Griffin Street
Dallas, TX 75202
(214) 767-6970

OREGON
Bureau of Labor Statistics
71 Stevenson Street
P.O. Box 3766
San Francisco, CA 94119
(415) 744-6600

PENNSYLVANIA
Bureau of Labor Statistics
3535 Market Street
P.O. Box 13309
Philadelphia, PA 19101
(215) 596-1154

PUERTO RICO
Bureau of Labor Statistics
Room 808
201 Varick Street
New York, NY 10014
(212) 337-2400

RHODE ISLAND
Bureau of Labor Statistics
1 Congress Street
10th Floor
Boston, MA 02114
(617) 565-2327

SOUTH CAROLINA
Bureau of Labor Statistics
1371 Peachtree Street N.E.
Atlanta, GA 30367
(404) 347-4416

SOUTH DAKOTA
Bureau of Labor Statistics
911 Walnut Street
Kansas City, MO 64106
(816) 426-2481

TENNESSEE
Bureau of Labor Statistics
1371 Peachtree Street N.E.
Atlanta, GA 30367
(404) 347-4416

TEXAS
Bureau of Labor Statistics
Room 22, Federal Building
525 Griffin Street
Dallas, TX 75202
(214) 767-6970

TRUST TERRITORY OF THE PACIFIC ISLANDS
Bureau of Labor Statistics
71 Stevenson Street
P.O. Box 3766
San Francisco, CA 94119
(415) 744-6600

UTAH
Bureau of Labor Statistics
911 Walnut Street
Kansas City, MO 64106
(816) 426-2481

VERMONT
Bureau of Labor Statistics
1 Congress Street, 10th Floor
Boston, MA 02114
(617) 565-2327

VIRGIN ISLANDS
Bureau of Labor Statistics
Room 808
201 Varick Street
New York, NY 10014
(212) 337-2400

VIRGINIA
Bureau of Labor Statistics
3535 Market Street
P.O. Box 13309
Philadelphia, PA 19101
(215) 596-1154

WASHINGTON
Bureau of Labor Statistics
71 Stevenson Street
P.O. Box 3766
San Francisco, CA 94119
(415) 744-6600

WEST VIRGINIA
Bureau of Labor Statistics
3535 Market Street
P.O. Box 13309
Philadelphia, PA 19101
(215) 596-1154

WISCONSIN
Bureau of Labor Statistics
9th Floor
Federal Office Building
230 S. Dearborn Street
Chicago, IL 60604
(312) 353-1880

WYOMING
Bureau of Labor Statistics
911 Walnut Street
Kansas City, MO 64106
(816) 426-2481

The National Agricultural Statistics Service
State Statisticians' Offices

The National Agricultural Statistics Service (NASS) has a staff of 45 state statisticians that serve all fifty states. They collect and distribute data for each state on **crops, livestock, poultry, dairy, prices, and labor.**

For details about agricultural reports issued by the NASS Agricultural Statistics Board call: **(202) 720-4020.**

For information about computerized data available through both the Agricultural Statistics Board and private computer networks call: **(202) 720-6306.**

National Agricultural Statistics Service State Statisticians' Offices

ALABAMA
State Statistician
Jewell T. Barr
P.O. Box 1071
Montgomery, AL 361-1071
(205) 223-7263

ALASKA
State Statistician
Delon A. Brown
P.O. Box 799
Palmer, AK 99645
(907) 745-4272

ARIZONA
State Statistician
Barry L. Bloyd
Suite 250
201 East Indianola
Phoenix, AZ 85012
(602) 640-2573

ARKANSAS
State Statistician
Benjamin F. Klugh
P.O. Box 3197
Little Rock, AR 72203
(501) 378-5145

CALIFORNIA
State Statistician
H. James Tippett
P.O. Box 1258
Sacramento, CA 95812
(916) 551-1533

COLORADO
State Statistician
Charles A. Hudson
P.O. Box 150969
Lakewood, CO 80215
(303) 236-2300

CONNECTICUT
State Statistician
Abrey R. Davis
P.O. Box 1444
Concord, NH 03302
(603) 224-9639

DELAWARE
State Statistician
Thomas W. Feurer
Delaware Department of Agriculture
2320 S. Dupont Highway
Dover, DE 19901
(302) 736-4811

FLORIDA
State Statistician
Robert L. Freie
1222 Woodward Street
Orlando, FL 32803
(407) 648-6013

GEORGIA
State Statistician
Larry E. Snipes
Suite 320
Stephens Federal Building
Athens, GA 30613
(404) 546-2236

HAWAII
State Statistician
Homer K. Rowley
P.O. Box 22159
Honolulu, HI 96823
(808) 973-9588

IDAHO
State Statistician
Donald G. Gerhardt
P.O. Box 1699
Boise, ID 83701
(208) 334-1507

ILLINOIS
State Statistician
Jerry L. Clampet
P.O. Box 19283
Springfield, IL 62794
(217) 492-4295

INDIANA
State Statistician
Ralph W. Gann
Room 223, 1148 Agad Building
Purdue University
West Lafayette, IN 47907
(317) 494-8371

IOWA
State Statistician
Duane M. Skow
833 Federal Building
210 Walnut Street
Des Moines, IA 50309
(515) 284-4340

KANSAS
State Statistician
Thomas J. Byram
P.O. Box 3534
Topeka, KS 66601
(913) 233-2230

KENTUCKY
State Statistician
David D. Williamson
P.O. Box 1120
Louisville, KY 40201
(502) 582-5293

LOUISIANA
State Statistician
Albert D. Frank
P.O. Box 65038
Baton Rouge, LA 70896
(504) 922-1362

MAINE
State Statistician
Abrey R. Davis
P.O. Box 1444
Concord, NH 03302
(603) 224-9639

MARYLAND
State Statistician
Melvin B. West
Suite 202
50 Harry Truman Parkway
Annapolis, MD 21401
(301) 841-5740

MASSACHUSSETTS
State Statistician
Abrey R. Davis
P.O. Box 1444
Concord, NH 03302
(603) 224-9639

MICHIGAN
State Statistician
Donald J. Fedewa
P.O. Box 20008
Lansing, MI 48901
(517) 377-1831

MINNESOTA
State Statistician
Carroll C. Rock
P.O. Box 7068
St. Paul, MN 55107
(612) 296-2230

MISSISSIPPI
State Statistician
George R. Knight
P.O. Box 980
Jackson, MS 39205
(601) 965-4575

MISSOURI
State Statistician
Paul A. Walsh
P.O. Box L
Columbia, MO 65205
(314) 876-0950

MONTANA
State Statistician
James Sands
P.O. Box 4369
Helena, MT 59604
(406) 449-5303

NEBRASKA
State Statistician
Jack L. Aschwege
P.O. Box 81069
Lincoln, NE 68501
(402) 437-5541

NEVADA
State Statistician
Clemence R. Lies
P.O. Box 8880
Reno, NV 89507
(702) 784-5584

NEW HAMPSHIRE
State Statistician
Abrey R. Davis
P.O. Box 1444
Concord, NH 03302
(603) 224-9639

NEW JERSEY
State Statistician
Robert J. Battaglia
Room 204
Health and Agriculture Building
CN-330 New Warren Street
Trenton, NJ 08625
(609) 292-6385

NEW MEXICO
State Statistician
Charles E. Gore
P.O. Box 1809
Las Cruces, NM 88004
(505) 522-6023

NEW YORK
State Statistician
Robert E. Schooley
Department of Agriculture
1 Winner Circle
Albany, NY 12235
(518) 457-5570

NORTH CAROLINA
State Statistician
Robert M. Murphy
P.O. Box 27767
Raleigh, NC 27611
(919) 856-4394

NORTH DAKOTA
State Statistician
Steve D. Wiyatt
P.O. Box 3166
Fargo, ND 58108
(701) 239-5306

OHIO
State Statistician
James E. Ramey
Room 608, New Federal Building
200 North High Street
Columbus, OH 43215
(614) 469-5590

OKLAHOMA
State Statistician
Robert P. Bellinghausen
2800 North Lincoln Boulevard
Oklahoma City, OK 73105
(405) 525-9226

OREGON
State Statistician
Paul M. Williamson
1735 Federal Building
1220 S.W. Third Avenue
Portland, OR 97204
(503) 326-2131

PENNSYLVANIA
State Statistician
Wally C. Evans
Room G-19
2301 N. Cameron Street
Harrisburg, PA 17110
(717) 787-3904

RHODE ISLAND
State Statistician
Abrey R. Davis
P.O. Box 1444
Concord, NH 03302
(603) 224-9639

SOUTH CAROLINA
State Statistician
Harry J. Power
P.O. Box 1911
Columbia, SC 29202
(803) 765-5333

SOUTH DAKOTA
State Statistician
John C. Ranek
P.O. Box 5068
Sioux Falls, SD 57117
(605) 330-4235

TENNESSEE
State Statistician
Charles R. Brantner
P.O. Box 41505
Nashville, TN 37204
(615) 781-5300

TEXAS
State Statistician
Dennis S. Findley
P.O. Box 70
Austin, TX 78767
(512) 482-5581

UTAH
State Statistician
Delroy J. Gneiting
P.O. Box 25007
Salt Lake City, UT 84125
(801) 524-5003

VERMONT
State Statistician
Abrey R. Davis
P.O. Box 1444
Concord, NH 03302
(603) 224-9639

VIRGINIA
State Statistician
Robert Bass
P.O. Box 1659
Richmond, VA 23213
(804) 786-3500

WASHINGTON
State Statistician
Douglas A. Hasslen
6128 Capitol Boulevard
Tumwater, WA 98501
(206) 586-8919

WEST VIRGINIA
State Statistician
Dave Loos
WV Department of Agriculture
Charleston, WV 25305
(304) 348-2217

WISCONSIN
Lyle H. Pratt
P.O. Box 9160
Madison, WI 53715
(608) 264-5317

WYOMING
State Statistician
Richard W. Coulter
P.O. Box 1148
Cheyenne, WY 82003
(307) 772-2181

Section III. Commercial Data Services

The Census Bureau maintains a list of businesses (including a few government and academic organizations) that have products and services incorporating Census Bureau data. These organizations are enrolled in the **Census Bureau's National Clearing House for Census Data Services**. Keep in mind that the Census Bureau does not endorse these organizations and makes no attempt to judge the quality of the products and services they supply. Since most economists and statisticians in the Data Factories know a surprising amount about any commercial data series, seek their advice as well.

Note—Most of the businesses and other organizations enrolled in the **National Clearing House for Census Data Services** also perform services using data produced by other government Data Factories.

Census Bureau National Clearing House Organizations

ALABAMA

Intergraph Corporation
Map Stop IW17A2
Huntsville, AL 35894
Andrew Weatherington
(205) 730-2000

Locational Data Systems, Inc.
309 Canal Street, NE
Decatur, AL 35601
James Skiles / Andy Kinney
(205) 340-1480

ARIZONA

Glimpse Econometrics
P.O. Box 5738
Scottsdale, AZ 85261
Richard Froncek
(602) 948-7688

IS Southwest
1245 E. Topeka Drive
Phoenix, AZ 85024
B.J. Raval
(602) 254-0977

CALIFORNIA

Advanced Technology Center
2298 Millcreek Drive
Laguna Hills, CA 92653
Larry Paulson
(714) 583-9119

Area Location Systems, Inc.
9410 Topanga Canyon Boulevard
Suite 110
Northridge, CA 91311
Mark Behnke
(818) 993-4275

Biddle and Associates, Inc.
903 Enterprise Drive, Suite 1
Sacramento, CA 95825
Cheryl Morgan
(800) 999-0438

Brighter Images, Inc.
936 Dewing Avenue, Suite J
Layfayette, CA 94549
Larry Fulcher
(510) 283-3340

Coast Meridian Marketing, Inc.
4029 Westerly Place, Suite 113
Newport Beach, CA 92660
Staff
(714) 752-8622

DATA QUICK
9171 Town Center Drive
Suite 600
San Diego, CA 92122
Lynn Sites
(619) 455-6900

Demographic Research Company
2221 Rosecrans Avenue, Suite 111
El Segundo, CA 90245
Joseph J. Weissmann
(213) 643-7588

Dynamic Ventures
992 Inverness Way
Sunnyvale, CA 94087
Myrna Ehrlich
(408) 732-7593

Educational Data Systems, Inc.
901 Campisi Way, Suite 160
Campbell, CA 95008
William Gilmore
(408) 559-4424

ESRI
380 New York Street
Redlands, CA 92373
Earl Nordstrand
(714) 793-2853

ETAK, Inc.
1430 O'Brien Drive
Menlo Park, CA 94025
William L. Folchi
(415) 328-3825

Equifax National Decisions
Systems
539 Encinitas Boulevard
Encinitas, CA 92024
Daniel Davies
(619) 942-7000

Expert Database Marketing Systems
15707 Rockfield Boulevard
Suite 250
Irvine, CA 92718
David Deeter
(714) 768-5775

Facility Mapping Systems, Inc.
38 Miller Avenue, Suite 11
Mill Valley, CA 94941
Dennis Klein / Lynne Finlay
(415) 381-1750

Foreign Trade Data Services
17527 Live Oak Circle
Fountain Valley, CA 92708
Don Dennison
(714) 964-9898

GEOSOFT Corporation
3547 Old Conejo Road
Suite 102
Newbury Park, CA 91320
Mike Anderson
(805) 499-2446

Klynas Engineering
P.O. Box 499
Simi Valley, CA 93062
Scott Klynas
(805) 529-1717

National Planning Data
Corporation
1801 Avenue of the Stars
Suite 729
Los Angeles, CA 90067
Mark Reiswig
(213) 557-0158

Nobi Takahashi and Associates
P.O. Box 1319
Oakland, CA 94604
Nobi Takahashi
(415) 465-0293

Recordata West, Inc.
2501 West Burbank Boulevard
Suite 202
Burbank, CA 91505
Edward Kasman
(818) 954-0132

Renaissance Automation
4455 Torrance Boulevard
Suite 342
Torrance, CA 90503
Charlie Szymanski
(310) 375-6922

STREET WISE
2910 Neilson Way, Suite 604
Santa Monica, CA 90405
Frank Hoeschler
(213) 452-1787

SciData Research, Inc.
3244 Camino Diablo
Lafayette, CA 94549
Albert Borden
(510) 939-1960

SourcePoint
401 B Street, Suite 800
San Diego, CA 92101
Eunice Tanjuaquio
(619) 595-5353

Strategic Mapping, Inc.
4030 Moorpark Avenue
Suite 250
San Jose, CA 95117
Lanning Forrest
(408) 985-7400

Thomas Bros. Maps
17931 Cowan
Ivine, CA 92714
Charles F. Cone
(714) 863-1984

Urban Decisions Systems, Inc.
2040 Armacost Avenue
Los Angeles, CA 90025
John Hobson
(213) 820-8931

Urban Microsystems
1305 Franklin Street, Suite 2001
Oakland, CA 94612
Pouilcos Prastacos
(415) 836-0804

Western Economic
Research Co., Inc.
8155 Van Nuys Boulevard
Suite 100
Panorama City, CA 91402
Michael Long
(818) 787-6277

COLORADO

GENASYS, Inc.
2629 Redwing Road, Suite 330
Fort Collins, CO 80526
Tom Bramble
(303) 226-3283

Generation 5 Technology, Inc.
8670 Wolff Court, Suite 200
Westminster, CO 80030
Rick Garfield
(303) 427-0055

GeoSpatial Solutions, Inc.
2450 Central Avenue
Suite E-1
Boulder, CO 80301
Caren McMahan
(303) 442-6622

GIS World, Inc.
2629 Redwing Road, Suite 280
Fort Collins, CO 80526
Derry Eynon
(303) 223-4848

Leica, Inc.
303 East 17th Avenue, Suite 440
Denver, CO 80112
Rob Van Westenberg
(303) 799-9453

Micro Map and CAD
9642 W. Virginia Circle
Lakewood, CO 80226
Randy George
(303) 988-4940

Precision Visuals, Inc.
6260 Lookout Road
Boulder, CO 80301
Chris Logan
(303) 530-9000

Public Systems Associates, Inc.
303 East 17th Avenue, Suite 440
Denver, CO 80203
Craig Butler
(303) 831-1260

SMARTSCAN, Inc.
2344 Spruce Street
Boulder, CO 80302
Rebecca Culp
(303) 443-7226

CONNECTICUT

Donnelley Marketing
Information Services
70 Seaview Avenue
Stamford, CT 06904
Jack Proehl
(203) 353-7295

Robert H. Frost
P.O. Box 495
Essex, CT 06426
Bob Frost
(203) 767-1254

Labtek Corporation
565 Wagon Trail
Orange, CT 06477
Thomas Griest
(203) 877-2880

DISTRICT OF COLUMBIA

Charles R. Mann Associates, Inc.
1828 L Street, N.W.
Washington, DC 20036
Charles R. Mann
(202) 466-6161

Colman Levin & Associates
2301 N Street, N.W., Suite 306
Washington, DC 20037
Colman Levin
(202) 223-0716

Election Data Services, Inc.
1225 I Street, N.W., Suite 700
Washington, DC 20005
Ronda Sternberg
(202) 789-2004

International Data and Development
2100 M Street, N.W., Suite 200
Washington, DC 20037
Staff
(202) 872-5245

National Safety Council
1019 19th Street, N.W.
Room 401
Washington, DC 20036
Staff
(202) 293-2270

Slater Hall Information Products
1522 K Street, N.W., Suite 522
Washington, DC 20005
George Hall
Courtney Slater
(202) 682-1350

System Dynamics, Inc.
409 12th Street, S.W.
Suite LL10
Washington, DC 20024
Mark Fisher Bryant
(202) 863-3840

U.S. Environmental Protection Agency
Chemical Emergency Preparedness & Prevention Office (OS-120)
Washington, DC 20460
Tony Jover / Melanie Hoff
(202) 260-5338

FLORIDA

Behavioral Science Research
2121 Ponce de Leon Boulevard
Suite 1250
Coral Gables, FL 33134
Robert A. Ladner
(305) 443-2000

ComGrafix, Inc.
920 E Street
Clearwater, FL 34616
Kerry Mitchell
(813) 443-6807

THG Publishing Company
P.O. Box 1621
St. Petersburg, FL 33731
Francis L. Hanigan

University of Florida Libraries
Census Access Program
Library West 148
Gainesville, FL 32611
Ray Jones
(904) 392-0361

CEMR, College of Business Administration
University of South Florida
4202 E. Fowler Avenue
Tampa, FL 33620
Thomas A. Charles
(813) 974-4266

GEORGIA

CSRA Regional Development
Center
P.O. Box 2800
Augusta, GA 30914
David Jenkins
(404) 737-1823

ERDAS, Inc.
2801 Buford Highway
Suite 300
Atlanta, GA 30329
Andrea Gernazian
(404) 248-9000

GEOVISION, Inc.
5680 Peachtree Parkway
Norcross, GA 30092
Kenneth S. Shain
(404) 448-8224

Lowe Engineers, Inc.
7100 Peachtree Dunwoody Road
Atlanta, GA 30328
Bill Bersson
(404) 399-6400

Montage Information Systems, Inc.
1650 Oakbrook Drive, Suite 435
Norcross, GA 30093
Ronald Lingerfelt
(404) 840-0183

Spatial Technologies, Inc.
430 10th Street, ATDC
Suite S 101
Atlanta, GA 30318
Joseph G. Jay
(404) 892-4780

IDAHO

Geographic General, Inc.
3350 Americana Terrace
Suite 320
Boise, ID 83706
Dave Spencer
(208) 343-1181

Idaho State University
Center for Business Research and
Services
Campus Box 8450
Pocatello, ID 83209
Paul Zelus
(208) 236-2504

ILLINOIS

Concordia College
7400 Augusta Street
River Forest, IL 60305
Peter M. Becker
(312) 771-8300

Decision Sciences, Inc.
9133 North Long, Suite 100
Skokie, IL 60077
Chuck Jones
(708) 965-1581

Management Graphics
233 East Wacker Drive
Suite 3011
Chicago, IL 60601
Robert L. Harris
(312) 819-0645

Manuel Plotkin Research and Planning
625 N. Michigan Avenue
Suite 500
Chicago, IL 60611
Manuel Plotkin
(312) 751-4270

Pivar Computing Services, Inc.
165 Arlington Heights Road
Buffalo Grove, IL 60089
Don Prosek / Gary Pivar
(708) 459-6010

Street Map Software
1014 Boston Circle
Schaumburg, IL 60193
Lynn Barton
(708) 529-4044

Universal Statistics, Inc.
7550 Plaza Court
Willowbrook, IL 60521
Roger Stanley
(708) 325-5555

INDIANA

Fisher and Associates
4355 E. Old Oyers Road
Bloomington, IN 47408
Stephen Fisher
(812) 339-5736

MSE Corporation
941 North Meridian Street
Indianapolis, IN 46204
Patricia Alebis
(317) 634-1000

KANSAS

Ruf Corporation
1533 E. Spruce
Olathe, KS 66061
Brian Ruf
(913) 782-8544

LOUISIANA

Synthesis, Inc.
10342 Mayfair Drive
Suite 10
Baton Rouge, LA 70809
Curtis Lee
(504) 291-4768

MAINE

DeLorme Mapping
Main Street
P.O. Box 298
Freeport, ME 04032
Ann Simonite
(207) 865-4171

KORK Systems, Inc.
81 Park Street
Bangor, ME 04401
Virginia Whitaker
(207) 945-6353

MARYLAND

Altek Corporation
12210 Plum Orchard Drive
Silver Spring, MD 20904
E.A. Cameron
(301) 572-2555

Business Resources Group, Inc.
7910 Longbranch Parkway
Takoma Park, MD 20912
Laurie Burch
(301) 961-7353

Caliper Corporation
4819 Cumberland Avenue
Chevy Chase, MD 20815
Howard Simkowitz
(301) 654-4704

Congressional Information Service
4520 East-West Highway
Suite 800
Bethesda, MD 20814
Sharon Schedicke
(301) 654-1550

Ed Nichols Associates
10400 Connecticut Avenue
Suite 604
Kensington, MD 20895
Ed Nichols
(301) 946-8212

GIP Corporation
One Clemson Court
Rockville, MD 20850
Al Tavakoli
(301) 217-0105

GeoVisual Business Products
12700 Virginia Manor Road
Beltsville, MD 20705
Schera Chadwick
(301) 470-0100

Greenhorne & O'Mara, Inc.
9001 Edmonston Road
Greenbelt, MD 20770
Staff
(301) 982-2853

Group 1 Software, Inc.
6404 Ivy Lane, Suite 500
Greenbelt, MD 20770
Alan Slater
(800) 368-5806

HALLIBURTON-NUS
Environmental
910 Clopper Road
Gaithersburg, MD 20878
Charles Gillies
(301) 258-2568

Ricercar, Inc.
6422 Dahlonega Road
Bethesda, MD 20816
Jonathan Robbin
(301) 229-1552

Roadnet Technologies, Inc.
2311 York Road
Timonium, MD 21093
Len Kennedy
(301) 560-0030

STX Remote Sensing Services
4400 Forbes Boulevard
Lanham, MD 20706
Richard Irish
(301) 794-5020

MASSACHUSETTS

ATLAS Data Systems
730 Boston Post Road
Sudbury, MA 01776
Louie Ming
(508) 443-4877

Analysis and Forecasting, Inc.
P.O. Box 415
Cambridge, MA 02138
John Pitkin
(617) 491-8171

Applied Insurance Research
264 Newbury Street
Boston, MA 02116
Mary Porter
(617) 267-6645

Caliper Corporation
1172 Beacon Street
Newton, MA 02161
Howard Slavin
(617) 527-4700

DARATECH, Inc.
140 Sixth Street
P.O. Box 410
Cambridge, MA 02142
Barbara Mende
(617) 354-2339

Geo Data Analytics, Inc.
19 Parker Street
Melrose, MA 02176
John Connery
(617) 665-8130

Harte-Hanks Data Technologies
25 Linnell Circle
Billerica, MA 01821
William Maxfield
(508) 667-7297

Intelligent Computer Engineering
One Business Way
Hopedal, MA 01747
Paul Desjourdy
(505) 478-4880

Market Planning Resources
85 Nowell Road
Melrose, MA 02176
Paul Landry
(617) 665-8589

Queues Enforth Development, Inc.
432 Columbia Street
Cambridge, MA 02141
George Fosque
(617) 225-2510

Schofield Brothers, Inc.
1071 Worcester Road
Framingham, MA 01701
Staff
(508) 879-0030

MICHIGAN

APB Associates, Inc.
17321 Telegraph, Suite 204
Detroit, MI 48219
Patricia C. Becker
(313) 535-2077

Aangstrom Precision Corp.
5805 E. Pickard, Suite 160
Mt. Pleasant, MI 48858
F. Bryan Davies
(517) 772-2232

Center for Remote Sensing
Michigan State University
302 Berkey Hall
East Lansing, MI 48824
William Enslin
(517) 353-7195

Interuniversity Consortium for
Political & Social Research
P.O. Box 1248
Ann Arbor, MI 48106
Erik W. Austin
(313) 763-5010

Manatron, Inc.
2970 S. 9th Street
Kalamazoo, MI 49009
Mark Kemper
(616) 375-5300

Southeast Michigan Council of
Governments
1990 Edison Plaza
600 Plaza Drive
Detroit, MI 48226
Staff
(313) 961-4266

MINNESOTA

DATAMap, Inc.
7525 Mitchell Road
Eden Prairie, MN 55344
Dianne Runnels
(800) 533-7742

Martinez Corporation
240 East Fillmore Avenue
P.O. Box 7023
St. Paul, MN 55107
Tony Martinez
(612) 291-1127

MISSISSIPPI

Mississippi State University
Dept. of Sociology and
Anthropology
P.O. Drawer C
Mississippi State, MS 39762
Mohamed El-Attar
(601) 325-7886

MISSOURI

East-West Gateway Coordinating
Council
911 Washington Avenue
St. Louis, MO 63101
Kathryn Mack
(312) 421-4220

M.J. Harden Associates
720 Troost Avenue
Kansas City, MO 64106
Kelly Cobb
(816) 842-0141

McDonnell Douglas Systems
Integrati
13736 Riverport Drive
Hazelwood, MO 63043
Mark Hollingsheads
(314) 344-4165

University of MO at St. Louis
Urban Information Center
8001 Natural Bridge Road
St. Louis, MO 63121
John G. Blodgett
(314) 553-6014

MONTANA

Education Logistics, Inc.
1024 South Avenue West
Missoula, MT 59801
Marie Quinto
(406) 728-0893

GeoResearch
115 N. Broadway
Billings, MT 59101
Darrel Peterson
(406) 248-6771

Logistics Systems, Inc.
1024 South Avenue West
Missoula, MT 59801
Jerry Schlesinger
(406) 728-0921

NEBRASKA

MicroImages, Inc.
201 North 8th Street, Suite 15
Lincoln, NE 68508
Lee Miller
(402) 477-9554

NEW JERSEY

GEOSTAT
P.O. Box K
Rocky Hill, NJ 08553
R.A.B. Sargeaunt
(609) 924-7177

Intelligent Charting, Inc.
600 International Drive
Mt. Olive, NJ 07828
Richard B. Miller
(201) 691-7000

NEW HAMPSHIRE

Geographic Data Technology, Inc.
13 Dartmouth College Highway
Lyme, NH 03768
Warren Whitney
(603) 795-2183

TerraLogics
114 Daniel Webster Highway
South, Suite 348
Nashua, NH 03060
Matthew Goldworm
(603) 889-1800

NEW YORK

American Demographics
P.O. Box 68
Ithaca, NY 14851
Staff
(607) 273-6343

City University of NY
CUNY Data Center
33 West 42nd Street
Room 1446
New York, NY 10036
Staff
(212) 642-2085

Earth Info Sciences, Inc.
241 Warner Road
Lancaster, NY 14086
Edward L. Moll
(716) 685-4230

Erie & Niagara Counties Regional
Planning Board
3103 Sheridan Drive
Amherst, NY 14226
Gary Smith
(716) 837-2035

Financial Marketing Group, Inc.
599 Lexington Avenue
Suite 2300
New York, NY 10022
Brandon Lee
(212) 754-7938

Fischer Associates
4 Larkin Drive
Ballston Lake, NY 12019
Kathleen Fischer
(518) 384-1102

Geo Demographics, Ltd.
69 Arch Street
Johnson City, NY 13790
Daniel Jardine / David Semo
(607) 729-5220

IBM Corporation
Neighborhood Road
MS 5933
Kingston, NY 12401
Brian Nolan
(914) 385-5063

M.A.P. Systems International
258 Broadway
Troy, NY 12180
Ron Schrimp
(518) 271-5135

MapInfo Corporation
200 Broadway
Troy, NY 12180
Austin Fisher
(518) 274-8673

Market Statistics
633 Third Avenue
New York, NY 10017
Edward J. Spar
(212) 984-2380

National Planning Data
Corporation
P.O. Box 610
Ithaca, NY 14851
John Belcher
(607) 273-8208

New York Transport Council
1 World Trade Center, 82E
New York, NY 10048
Juliette Bergman
(212) 938-3352

Roger Creighton Associates
274 Delaware Avenue
Delmar, NY 12054
Brant Gardner
(518) 439-4991

Sanborn Mapping and Geographic
Information Services
629 Fifth Avenue
Pelham, NY 10803
Allan Davis
(914) 738-1649

Space Track, Inc.
75 Spring Street, 8th Floor
New York, NY 10012
John Ziegler
(212) 226-0522

Specialists in Business Information
3375 Park Avenue, Suite 2000A
Wantaugh, NY 11793
Stuart Hirschhorn
(516) 781-4934

NORTH CAROLINA

ATG Incorporated
205 Regency Executive Park
Suite 306
Charlotte, NC 28217
Ed Campbell
(704) 521-8113

Good Deals
310 Kingston Road
Knightdale, NC 27545
Clark Trivett
(919) 733-3809

INFOCEL, Inc.
4800 Six Forks Road
Raleigh, NC 27609
Steve Lindsay
(919) 783-8000

Pinnacle Graphics Software
208 Forsyth Drive
P.O. Box 3381
Chapel Hill, NC 27514
Stephen Smith
(919) 929-8013

SAS Institute, Inc.
SAS Campus Drive
Cary, NC 27513
John McIntyre

Westvaco
309 N. Channel Drive
Wrightsville Beach, NC 28480
Carolyn Souther
(914) 256-0048

OHIO

NODIS
Cleveland State University
College of Urban Affairs
Cleveland, OH 44115
Mark Salling
(216) 687-2209

Northeast Ohio Areawide
Coordinating Agency
668 Euclid Avenue
Cleveland, OH 44114
Staff
(216) 291-2414

Woolpert Geographic Information
Services
409 E. Monument Avenue
Dayton, OH 45402
Rex Cowden
(513) 461-5660

PENNSYLVANIA

BonData
245 West High Street
Hummelstown, PA 17036
Lisa Bontempo
(717) 566-5550

Geo Decisions, Inc.
118 Boalsburg Road
P.O. Box 1028
Lemont, PA 16851
Chris Markel
(814) 234-8625

GEOGRAPHIX, Inc.
156 North 3rd Street
Philadelphia, PA 19106
Roger Prichard
(215) 925-6690

Help Business Services, Inc.
HBS Building
110 Park Avenue
Swarthmore, PA 19081
John R. Kaufman
(215) 544-9787

Institute for Research and
Community Service
Indiana University of Pennsylvania
Indiana, PA 15705
Robert Sechrist
(412) 357-2251

Keystone Management Systems
522 E. College Avenue, Suite 200
State College, PA 16801
Gil Boettcher
(814) 234-6264

Michael Baker, Jr., Inc.
4301 Dutch Ridge Road
Box 280
Beaver, PA 15009
John Ferketic
(412) 495-4025

TEXAS

Contemporary Technological
Corporation
3701 West Alabama, Suite 460
Houston, TX 77027
Bernie Peterson
713) 621-8166

Conversion Resources Corporation
1802 NE Loop 410, Suite 500
San Antonio, TX 78217
Robert W. Thompson
(512) 829-7253

IBM Corporation
3700 Bay Area Boulevard
Mc 8126
Houston, TX 77058
Robert L. Gard
(713) 335-3201

Map Resources, Inc.
208 West 14th Street
Austin, TX 78701
Richard Hair
(512) 476-3113

National Planning Data
Corporation
14679 Midway Road, Suite 221
Dallas, TX 75244
Staff
(214) 980-0198

North Central Texas Council of
Governments
P.O. Drawer COG
Arlington, TX 76005
Bob O'Neal
(817) 640-3300

Synercom Technology, Inc.
2500 City West Blvd, Suite 1100
Houston, TX 77042
Pat Hansen / Jodi Loyd
(713) 954-7000

ZYCOR, Inc.
220 Foremost Drive
Austin, TX 78745
Robert Brown
(512) 282-6699

VIRGINIA

Anderson & Associates, Inc.
100 Ardmore Street
Blacksburg, VA 24060
S.K. Anderson
(703) 552-5592

CACI Marketing Systems
9302 Lee Highway, Suite 310
Fairfax, VA 22031
Gary Madison
(703) 218-4400

Chadwyck-Healy, Inc.
1101 King Street
Alexandria, VA 22314
Michael Fischer
(703) 683-4890

Claritas Corporation (Headquarters)
201 N. Union Street
Alexandria, VA 22314
Staff
(703) 683-8300

GIS Corporation
8000 Tower Crescent Drive
Suite 820
Vienna, VA 22182
Said Khosrowshahi
(703) 761-6140

Public Data Resources
Virginia Commonwealth
University
P.O. Box 2008
Richmond, VA 23284
Robert D. Rugg
(804) 367-1134

Spatial Data Sciences, Inc.
8200 Greensboro Drive
Suite 1020
McLean, VA 22102
John Turner
(703) 893-0183

Tidewater Consultants, Inc.
160 Newton Road, Suite 401
Virginia Beach, VA 23462
J.C. Barenti
(804) 497-8951

U. S. Statistics
1101 King Street, Suite 601
Alexandria, VA 22314
Warren Glimpse
(703) 979-9699

Vigyan Inc.
5203 Leesburg Pike, Suite 900
Falls Church, VA 22041
Mike Paquette
(703) 931-1100

WASHINGTON

Gambrell Urban, Inc.
GIS Division
900 4th Avenue, Suite 1206
Seattle, WA 98164
John Schloser
(206) 467-6900

Geographic Technology, Inc.
335 Telegraph Road
Bellingham, WA 98226
Oswin Slade
(206) 734-5993

National Oceanic and Atmospheric
Administration
NOAA/OMA34
7600 Sand Point Way, NE
Seattle, WA 98115
Mark Miller
(206) 526-6317

Sammamish Data Systems, Inc.
1813 130th Avenue NE
Suite 216
Bellevue, WA 98005
Richard Schweitzer
(206) 867-1485

Star Software, Inc.
8541 Southeast 68th Street
Mercer Island, WA 98040
Pete Gallus
(206) 232-8021

WISCONSIN

American Digital Cartography
715 West Parkway Boulevard
Appleton, WI 54914
Michael Bauer
(414) 733-6678

GEOCODE, Inc.
2816 London Road, Suite 5
Eau Claire, WI 54701
Michael A. Hines
(715) 834-5058

Geographic Systems Corp.
504 North Adams Street
Green Bay, WI 54301
Judith Keneklis
(414) 433-1706

CANADA

GIRO, Inc.
1100 Cremazie Boulevard East
Suite 300
Montreal, Quebec H2P2X2
Nigel Hamer
(514) 374-9221

M31 Systems, Inc.
1111 St. Charles Street West
Suite 115 West Tower
Lonqueuil, Quebec J4K5G4
Staff
(514) 928-4600

XII. Environmental Data Directory

Environmental Data Directory

The environment has taken center stage as a global concern and it is fast becoming an important issue in the business community. Environmental policies, linked as they are to energy strategies and business considerations—**plant costs, technology investment and manufacturing options**—are going to require more careful analysis in the future. This directory fills the need for a practical guide to business-relevant environmental data.

The Environmental Data Directory focuses in on the most frequently used business-related environmental data. It may not cover everything, but we know that it covers the key areas that touch directly on business considerations. And, like the other directories in this volume, even if it doesn't identify the exact person or topic you specifically may need, it will give you the connection you need to get the data you want.

The Environmental Data Directory is focused on frequently sought-after environmental statistics—**toxic waste, emissions, wetlands, forests, herbicides, landfills, hazardous chemicals, pesticides, air quality**—and **much more.** In the Directory, we have identified over 300 of the most important data topics and the government experts with environmental data. The directory includes data contacts regarding the impact of the environment on the health of the general population and also identifies the key people at the national laboratories researching the environment.

Part Two: The DATAPHONER

National and International Environmental Data Sources

The best published source for information on national level environmental data is the **Guide to Selected Environmental Statistics in the U.S. Government.** It is one of the first major products developed by EPA's new Center for Environmental Statistics. It is a guide to environmental statistical programs and the primary and secondary statistics they generate.

For a copy of the Guide, call **(202) 260-2680**, or write to:

Office of Policy, Planning and Evaluation
Center for Environmental Statistics
U.S. Environmental Protection Agency (PM-223)
401 M Street, S.W.
Washington, DC 20460

The best source for information on international environmental data is the **INFOTERRA Network.** INFOTERRA is the International Environmental Information Exchange Network coordinated by the United Nations Environment Program. The INFOTERRA network comprises a partnership of 140 countries which has designated national focal points to promote the exchange of environmental information. Each country has agreed to provide environmental information to international requestors free or at a minimal charge.

To obtain the address of the INFOTERRA national focal point in a specific country,
call **(254 2) 230800**, fax (254 2) 226890, or write to:

Dr.Woyen Lee, Director
INFOTERRA Programme Activity Centre
United Nations Environment Programme
P.O. Box 30552
Nairobi, Kenya

In the United States, the INFOTERRA national focal point for all international data requests is the U.S. Environmental Protection Agency. Call **(202) 260-5917,** fax (202) 260-3923, or write to:

Linda Spencer, Manager
INFOTERRA/USA
U.S. Environmental Protection Agency (PM-211A)
401 M Street, S.W.
Washington, DC 20460

Environmental Data Directory

Major Categories

Air Quality

Land Use

Solid and Toxic Waste

Water Quality and Water Use

Other Policy Relevant Environmental Data

Part Two: The DATAPHONER

Organizational Index and Acronyms

AEERL	Air and Energy Engineering Research Laboratory, U.S. Environmental Protection Agency
ANL	Argonne National Laboratory, U.S. Department of Energy
AREAL	Atmospheric Research and Exposure Assessment Laboratory, U.S. Environmental Protection Agency
ATSDR	Agency for Toxic Substances and Disease Registry
BLM	Bureau of Land Management, U.S. Department of the Interior
BNL	Brookhaven National Laboratory
BOM	Bureau of Mines, U.S. Department of the Interior
CCIW	Canada Centre for Inland Water
CENSUS	Bureau of the Census, U.S. Department of Commerce
CERL	Corvallis Environmental Research Laboratory, U.S. Environmental Protection Agency
CES	Center for Environmental Statistics, U.S. Environmental Protection Agency
CG	Coast Guard, U.S. Department of Transportation
CTI	Center for Transportation Information, U.S. Department of Transportation
ERL	Environmental Research Laboratory, U.S. Environmental Protection Agency
ERS	Economic Research Service, U.S. Department of Agriculture
FDA	Food and Drug Administration, U.S. Department of Health and Human Services
FHA	Federal Highway Administration, U.S. Department of Transportation
FS	U.S. Forest Service, U.S. Department of Agriculture
FSIS	Food Safety and Inspection Service, U.S. Department of Agriculture
FWS	U.S. Fish and Wildlife Service, U.S. Department of the Interior
GS	Geological Survey, U.S. Department of the Interior
NAREL	National Air and Radiation Environmental Laboratory, U.S. Environmental Protection Agency
NASS	National Agricultural Statistics Service, U.S. Department of Agriculture
NCDC	National Climate Data Center, NOAA, U.S. Department of Commerce
NCHS	National Center for Health Statistics, U.S. Department of Health and Human Services
NMFS	National Marine Fisheries Service, NOAA, U.S. Department of Commerce
NOAA	National Oceanic and Atmospheric Administration, U.S. Department of Commerce
NOS	National Ocean Service, NOAA, U.S. Department of Commerce
NPS	National Park Service, U.S. Department of the Interior

XII. Environmental Data Directory

OAQPS	Office of Air Quality Planning and Standards, U.S. Environmental Protection Agency
OCRWM	Office of Civilian Radioactive Waste Management, U.S. Department of Energy
OERR	Office of Emergency and Remedial Response (Superfund), U.S. Environmental Protection Agency
OGDW	Office of Groundwater and Drinking Water, U.S. Environmental Protection Agency
OMS	Office of Mobile Sources, U.S. Environmental Protection Agency
OPP	Office of Pesticide Programs, U.S. Environmental Protection Agency
ORD	Office of Research and Development, U.S. Environmental Protection Agency
ORNL	Oak Ridge National Laboratory, U.S. Department of Energy
OSW	Office of Solid Waste, U.S. Environmental Protection Agency
OTS	Office of Toxic Substances, U.S. Environmental Protection Agency
OW	Office of Water, U.S. Environmental Protection Agency
RFF	Resources for the Future
SCS	Soil Conservation Service, U.S. Department of Agriculture

Air Quality XII. *Environmental Data Directory*

Air Quality

Subject	Source	Data Contact	Telephone
Acid Deposition, Causes and Effects	AEERL	Wagner, Janice	919-541-1818
Acid Precipitation, Status and Trends	AREAL	Reagan, James	919-541-4486
Acid Rain, Monitoring Project (ANC)	CERL	Stoddard, John	503-757-4427
Acid Rain, Monitoring Surface Waters	CERL	Stoddard, John	503-757-4427
Acid Rain, Monitoring Network	GS	Gibson, J. H.	303-491-1978
Air Quality, Global (U.S. Contact)	AREAL	Evans, Gardner	919-541-3887
Ambient Air Quality, Trends and Analysis	OAQPS	Curran, Thomas	919-541-5558
Ambient Concentrations of Air Pollutants	OAQPS	Summers, Jacob	919-541-5695
Carbon Dioxide Gas Related Data	ORNL	Jones, Sonja	615-574-0390
Carbon Monoxide and Lead Emissions	OAQPS	Curran, Thomas	919-541-5558
Climatic and Atmospheric Data	NCDC	Staff	704-259-0682
Climatic Variables	NCDC	Staff	704-259-0682
Electric Utilities, Emissions	ANL	Miller, Don	708-972-3946
Emission Sources and Facility Data	AEERL	Wagner, Janice	919-541-1818
Emissions and Compliance Data	OAQPS	Isbell, Chuck	919-541-5448
Emissions, Carbon Monoxide and Lead	OAQPS	Curran, Thomas	919-541-5558
Emissions, Commercial Fuel Consumption	ANL	Miller, Don	708-972-3946
Emissions, Current Trends and Estimates	ANL	Miller, Don	708-972-3946
Emissions, Electric Utilities	ANL	Miller, Don	708-972-3946
Emissions from Industrial Processes	ANL	Miller, Don	708-972-3946
Emissions from Transportation	ANL	Miller, Don	708-972-3946
Emissions from Motor Vehicles	OMS	Bontekoe, Eldert	313-668-4200
Emissions, Industrial Fuel Consumption	ANL	Miller, Don	708-972-3946
Emissions, Organic Compounds	OAQPS	Curran, Thomas	919-541-5558
Emissions, Particulates	OAQPS	Curran, Thomas	919-541-5558
Emissions, Residential Fuel Consumption	ANL	Miller, Don	708-972-3946
Emissions, Stationary Sources	OAQPS	Curran, Thomas	919-541-5558
Emissions, Stationary and Mobile Sources	ANL	Kohout, Edward	708-972-7644
Emissions, Sulfur Oxides	OAQPS	Curran, Thomas	919-541-5558
Emissions, Nitrogen Oxides	OAQPS	Curran, Thomas	919-541-5558
Fuel Combustion Emissions	OAQPS	Curran, Thomas	919-541-5558
Greenhouse Gas Related Data	ORNL	Jones, Sonja	615-574-0390
Industrial Fuel Consumption, Emissions	ANL	Miller, Don	708-972-3946
Industrial Processes, Emissions	ANL	Miller, Don	708-972-3946
Industrial Processes Generated Emissions	OAQPS	Curran, Thomas	919-541-5558
National Parks, Atmospheric Conditions	NPS	Malm, William	303-491-8292

Subject	Source	Data Contact	Telephone
National Parks, Air Quality Data	NPS	Carson, Bob	303-969-2072
Ozone Concentration Levels by County	BNL	Coveney, Elizabeth	516-282-2259
Particulates, Emission Estimates	OAQPS	Curran, Thomas	919-541-5558
Pollutants, Air and Meteorological Data	OAQPS	Summers, Jacob	919-541-5695
Pollution Abatement, Operating Costs	Census	Shapiro, Janet	301-763-1755
Pollution Abatement, Capital Expenditures	Census	Shapiro, Janet	301-763-1755
Pollution Abatement Activities in Mfg.	Census	Shapiro, Janet	301-763-1755
Pollution Emissions at Individual Facilities	OAQPS	Isbell, Chuck	919-541-5448
Pollution Levels in the Air	OAQPS	Curran, Thomas	919-541-5558
Precipitation Chemistry, Monitoring	GS	Pickering, Ranard	703-648-6875
Quality of Air in National Parks	NPS	Carson, Bob	303-969-2072
Radiation Levels in Air	NAREL	Luster, Geraldine	205-270-3433
Radiation Levels in Air	NAREL	Goode, Paula	205-270-3433
Radioactive Emissions from Nuclear Plants	BNL	Tichler, Joyce	516-282-3801
Radioactivity Associated with Air	NAREL	Petko, Charles	205-270-3400
Reactive Volatile Organic Compounds, Emissions	OAQPS	Curran, Thomas	919-541-5558
Releases of Radioactive Materials	BNL	Tichler, Joyce	516-282-3801
Solar Radiation	NCDC	Staff	704-259-0682
Solid Waste Disposal, Ambient Air Emissions	OAQPS	Curran, Thomas	919-541-5558
Studies of Air Quality in Metro. Areas	AREAL	Lawless, Thomas	919-541-2291
Sulfur Oxides, Emission Estimates	OAQPS	Curran, Thomas	919-541-5558
Toxic Air Pollutants	OAQPS	Pope, Anne	919-541-5373
Toxic Data (Air), National Clearinghouse	OAQPS	Kilaru, Vasu	919-541-0850
Toxic Releases, Atmospheric Emissions	OTS	Sasnett, Samuel	202-260-1821
Transportation, Emissions	ANL	Miller, Don	708-972-3946
Transportation Generated Emissions	OAQPS	Curran, Thomas	919-541-5558

Land Use

Subject	Source	Data Contact	Telephone
Acid Deposition, Causes and Effects	AEERL	Wagner, Janice	919-541-1818
Agricultural Chemicals, Amounts Used	NASS	Rives, Sam	202-720-2324
Agricultural Land, Acreages and Uses	ERS	Krupa, Ken	202-219-0424
Agricultural Land, Acreages and Uses	ERS	Daugherty, Arthur	202-219-0424
Agroecosystems, Ecological Status (EMAP)	ORD	Dixon, Thomas	202-260-5782
Arid Lands, Ecological Condition (EMAP)	ORD	Dixon, Thomas	202-260-5782
Cropland, Estimated Acreages and Major Uses	ERS	Krupa, Ken	202-219-0424
Cropland, Estimated Acreages and Major Uses	ERS	Daugherty, Arthur	202-219-0424
Ecological Condition of Forests (EMAP)	ORD	Van Remortel, Rick	702-734-3295
Emission Sources and Facility Data	AEERL	Wagner, Janice	919-541-1818
Erosion Data	SCS	George, Tommy	202-720-6267
Farm Land, Acres Irrigated	Census	Peterson, Dave	301-763-8560
Fires on Public and Private Wildlands	FS	Staff	202-205-1498
Fish and Wildlife Service Lands	FWS	Short, Olivia	703-358-1811
Forest Land, Acreages and Uses	ERS	Daugherty, Arthur	202-219-0424
Forest Lands, Insect and Disease Conditions	FS	Hofacker, Thomas	202-205-1600
Forest Resources, Inventory and Analysis	FS	Bones, James	202-205-1343
Forest System, U.S., Uses and Conditions	FS	Cron, Robert	202-205-1408
Forest Utilization, Public Lands Data	BLM	Ratliff, Michael	202-208-5717
Forests, Ecological Condition (EMAP)	ORD	Van Remortel, Rick	702-734-3295
Forests, National, Data on Grazing	FS	Williamson, Robert	202-205-1460
Forests, National, General Characteristics	FS	Dunning, Philip	202-205-0843
Grassland Pasture and Range, Acreages and Uses	ERS	Daugherty, Arthur	202-219-0424
Grassland Pasture and Range, Acreages and Uses	ERS	Krupa, Ken	202-219-0424
Grasslands, National, Data on Grazing	FS	Williamson, Robert	202-205-1460
Grazing Permits and Leases, Public Lands Data	BLM	Ratliff, Michael	202-208-5717
Hazardous Waste Sites, Abandoned/Uncontrolled	OERR	Staff	202-260-3770
Hazardous Waste Sites, Contaminants	ATSDR	Perry, Mike	404-639-0720
Hazardous Waste Sites Data (HAZDAT)	ATSDR	Perry, Mike	404-639-0720
Hazardous Waste Sites, Populations Impacted	ATSDR	Perry, Mike	404-639-0720
Hazardous Waste Sites, Site Characteristics	ATSDR	Perry, Mike	404-639-0720
Herbicide Use in Agricultural Crop Production	RFF	Gianessi, Leonard	202-328-5036
Highway Statistics, Characteristics and Uses	FHA	Jarema, Frank	202-366-0160
Hydrologic Hazards and Land Use	GS	Paulson, Richard	703-648-6851
Insects and Disease Conditions on Forest Lands	FS	Hofacker, Thomas	202-205-1600
Irrigation Data, Farm and Ranch Lands	Census	Peterson, Dave	301-763-8560

Part Two: The DATAPHONER   Land Use

Subject	Source	Data Contact	Telephone
Land Cover and Land Use	SCS	George, Tommy	202-720-6267
Land Cover Maps (Vegetation, Water, Etc.)	GS	Staff	703-860-6045
Land UseMaps	GS	Staff	703-860-6045
Landfills, Municipal, Hydrological Data	OSW	Galbreath, Myra	202-260-4697
Landfills, Municipal, Persons Using Facilities	OSW	Galbreath, Myra	202-260-4697
Landfills, Municipal, Quantities Generated	OSW	Galbreath, Myra	202-260-4697
Landfills, Municipal, Size and Capacity	OSW	Galbreath, Myra	202-260-4697
Landfills, Municipal, Types of Waste Refused	OSW	Galbreath, Myra	202-260-4697
Landfills, Municipal, Types of Waste Accepted	OSW	Galbreath, Myra	202-260-4697
Landfills,, Proximity to Drinking Water	OSW	Galbreath, Myra	202-260-4697
Lands Under Control of the U.S. FWS	FWS	Hawkins, Thomas	703-358-1811
Maps, Land Use Related to Human Activities	GS	Staff	703-860-6045
Mine Waste Statistics	BOM	Rogich, Donald	202-634-1187
Mineral and Surface Ownership, Public Lands	BLM	Ratliff, Michael	202-208-5717
National Park Lands, Characteristics	NPS	Minnick, Renee	202-343-3862
National Parks, Air Quality Data	NPS	Carson, Bob	303-969-2072
National Parks, Atmospheric Conditions	NPS	Malm, William	303-491-8292
Nonagricultural Land, Acreages and Major Uses	ERS	Krupa, Ken	202-219-0424
Nonagricultural Land, Acreages and Major Uses	ERS	Daugherty, Arthur	202-219-0424
Pesticides and Fertilizer, Amounts Used	NASS	Rives, Sam	202-720-2324
Public Lands Statistics	BLM	Ratliff, Michael	202-208-5717
Ranch Land, Acres Irrigated	Census	Peterson, Dave	301-763-8560
Range and Related Lands in National Forests	FS	Dunning, Philip	202-205-0843
Range Conditions, Public Lands Data	BLM	Ratliff, Michael	202-208-5717
Rangeland Data and Maps on Habitat Types	BLM	Staff	202-653-9193
Rangeland Resources, Inventory and Analysis	FS	Bones, James	202-205-1343
Recreation Sites in the U.S. Forest System	FS	Cron, Robert	202-205-1408
Soil Characteristics	SCS	George, Tommy	202-720-6267
Soil, Water and Related Resources Data	SCS	George, Tommy	202-720-6267
Timber Data, Acreage, Volume and Value	BLM	Bird, Dick	202-653-8864
Timber Sales	BLM	Bird, Dick	202-653-8864
Transportation Data, Energy Use	CTI	Bradley, Kathleen	617-494-2614
Treatment Data (Irrigation, Tillage, Etc.)	SCS	George, Tommy	202-720-6267
Tree Planting, Number of Seedlings	FS	Mangold, Robert	202-205-1379
Tree Planting, Number of Acres Planted	FS	Mangold, Robert	202-205-1379
Vegetative Conditions (Wetlands, Etc.)	SCS	George, Tommy	202-720-6267
Wetland Resources, Characteristics and Extent	FWS	Wilen, Bill	703-358-2201
Wetland Resources, Characteristics and Extent	FWS	Dahl, Thomas	813-893-3873
Wetland Resources, Status and Trends	FWS	Dahl, Thomas	813-893-3873
Wetland Resources, Status and Trends	FWS	Wilen, Bill	703-358-2201
Wetlands, Ecological Condition (EMAP)	ORD	Dixon, Thomas	202-260-5782

Land Use

Subject	Source	Data Contact	Telephone
Wetlands, National Inventory Maps	FWS	Staff	800USAMAPS
Wetlands, Vegetative Conditions	SCS	George, Tommy	202-720-6267
Wildlands, Public and Private, Fire Data	FS	Staff	202-205-1498

Solid and Toxic Waste

Subject	Source	Data Contact	Telephone
CERCLIS	OERR	Staff	202-260-3770
Hazardous Chemicals Data	OTS	Leitzke, John	202-260-3507
Hazardous Waste, Amount of Waste Generated	OSW	Galbreath, Myra	202-260-4697
Hazardous Waste, Amount of Waste Disposed	OSW	Galbreath, Myra	202-260-4697
Hazardous Waste, Disposal / Recycling Facilities	OSW	Galbreath, Myra	202-260-4697
Hazardous Waste, Management Facilities	OSW	Galbreath, Myra	202-260-4697
Hazardous Waste, Number of Generators	OSW	Galbreath, Myra	202-260-4697
Hazardous Waste Sites, Contaminants	ATSDR	Perry, Mike	404-639-0720
Hazardous Waste Sites Data (HAZDAT)	ATSDR	Perry, Mike	404-639-0720
Hazardous Waste Sites, Populations Impacted	ATSDR	Perry, Mike	404-639-0720
Hazardous Waste Sites, Site Characteristics	ATSDR	Perry, Mike	404-639-0720
Hazardous Waste Statistics	OSW	Galbreath, Myra	202-260-4697
Hazardous Waste, Treatment and Storage	OSW	Galbreath, Myra	202-260-4697
Landfills, Municipal, Hydrological Data	OSW	Galbreath, Myra	202-260-4697
Landfills, Municipal, Persons Using Facilities	OSW	Galbreath, Myra	202-260-4697
Landfills, Municipal, Quantities Generated	OSW	Galbreath, Myra	202-260-4697
Landfills, Municipal, Size and Capacity	OSW	Galbreath, Myra	202-260-4697
Landfills, Municipal, Types of Waste Refused	OSW	Galbreath, Myra	202-260-4697
Landfills, Municipal, Types of Waste Accepted	OSW	Galbreath, Myra	202-260-4697
Landfills, Proximity to Drinking Water	OSW	Galbreath, Myra	202-260-4697
Nonhazardous Waste by Type of Industry	OSW	Galbreath, Myra	202-260-4697
Nonhazardous Waste Disposal and Recycling	OSW	Galbreath, Myra	202-260-4697
Nonhazardous Waste Generated and Managed	OSW	Galbreath, Myra	202-260-4697
Nonhazardous Waste Treatment and Storage	OSW	Galbreath, Myra	202-260-4697
Pollution Abatement Activities in Mfg.	Census	Shapiro, Janet	301-763-1755
Pollution Abatement, Capital Expenditures	Census	Shapiro, Janet	301-763-1755
Pollution Abatement, Operating Costs	Census	Shapiro, Janet	301-763-1755
Radioactive Solid Waste Release	BNL	Tichler, Joyce	516-282-3801
Radioactive Waste and Spent Fuel	ORNL	Klein, Jerry	615-574-6823
Radioactive Waste and Spent Fuel Inventory	OCRWM	Payton, M. L.	202-586-9140
RCRA Data	OTS	Sasnett, Samuel	202-260-1821
Releases of Radioactive Materials	BNL	Tichler, Joyce	516-282-3801
Solid Waste Management by Government	Census	Wulf, Henry	301-763-7664
Toxic Releases, Land Disposal	OTS	Sasnett, Samuel	202-260-1821
Toxic Releases, Treatment Processes	OTS	Sasnett, Samuel	202-260-1821
Toxic Waste Reduction Data	OTS	Sasnett, Samuel	202-260-1821

Water Quality and Water Use

Subject	Source	Data Contact	Telephone
Acid Deposition, Causes and Adverse Effects	AEERL	Wagner, Janice	919-541-1818
Acid Precipitation, Status and Trends	AREAL	Reagan, James	919-541-4486
Acid Rain, Monitoring Project (ANC)	CERL	Stoddard, John	503-757-4427
Aquatic Toxicity Data	ERL	Pilli, Anne	218-720-5516
Availability, Potential Uses and Development	GS	Paulson, Richard	703-648-6851
Bacterial Water Quality, Shellfishing Waters	NOS	Slaughter, Eric	301-443-8843
Biological Characteristics of Lakes	ERL	Landers, Dixon	503-757-4427
Chemical Characteristics of Lakes	ERL	Landers, Dixon	503-757-4427
Coastal Areas, Status of Environmental Quality	NOS	O'Connor, Thomas	301-443-8644
Coastal Areas, Water Pollutant Discharges	OS	Farrow, Daniel	301-443-0454
Coastal Pollutant Discharge Estimates	NOS	Farrow, Daniel	301-443-0454
Coastal Waters and Bay Bottoms	FWS	Wilen, Bill	703-358-2201
Coastal Waters, Environmental Quality	NOS	O'Connor, Thomas	301-443-8644
Coastal Zone, Pollution Incidents and Oil Spills	CG	Staff	202-267-6993
Commercial Water Use	GS	Holmes, Sandra	703-648-6815
Contaminant Concentrations in Sediments	NOS	O'Connor, Thomas	301-443-8644
Contaminant in Marine Organisms	NOS	O'Connor, Thomas	301-443-8644
Contaminant Levels, Drinking Water Supplies	OGDW	Sexton, Cecil	202-260-7276
Contaminants Threatening Fish and Wildlife	FWS	Schmitt, Chris	314-875-1800
Domestic Water Use	GS	Holmes, Sandra	703-648-6815
Drinking Water, Radiation Levels	NAREL	Goode, Paula	205-270-3433
Drinking Water, Radiation Levels	NAREL	Luster, Geraldine	205-270-3433
Drinking Water Supplies, Contaminant Levels	OGDW	Sexton, Cecil	202-260-7276
Drinking Water Wells, Nitrate Levels	OPP	Staff	800-426-4791
Drinking Water Wells, Pesticide Concentrations	OPP	Staff	800-426-4791
Ecological Status in Coastal Waters (EMAP)	ORD	Latimer, Richard	401-782-3077
Ecological Status, Inland Waters (EMAP)	ORD	Latimer, Richard	401-782-3077
Ecological Status of Wetlands (EMAP)	ORD	Dixon, Thomas	202-260-5782
Emission Sources, Facility Data	AEERL	Wagner, Janice	919-541-1818
Estuarine Areas, Environmental Quality	NOS	O'Connor, Thomas	301-443-8644
Estuarine Waters, Environmental Quality	NOS	O'Connor, Thomas	301-443-8644
Fish and Wildlife, Contaminant Threats	FWS	Steffeck, Donald	703-358-2148
Fisheries, Annual World Catch Data	NMFS	Holliday, Mark	301-713-2328
Fisheries, Commercial Landings (Catch)	NMFS	Holliday, Mark	301-713-2328
Fisheries, Number of Vessels and Fishermen	NMFS	Holliday, Mark	301-713-2328
Fisheries, Processed Products Data	NMFS	Holliday, Mark	301-713-2328

Subject	Source	Data Contact	Telephone
Fisheries, Shrimp Imports by Country	NMFS	Holliday, Mark	301-713-2328
Groundwater, Chemical Characteristics	GS	Alexander, Richard	703-648-6869
Groundwater Level Data from 200 Sites	GS	Ross, Thomas	703-648-6814
Groundwater, Physical Characteristics	GS	Smith, Richard	703-648-6870
Groundwater, Quantity, Quality, Distribution	GS	Briggs, John	703-648-5624
Hazardous Substances, Spills in U.S. Waters	CG	Staff	202-267-6993
Hydroelectric Power Generation, Water Use	GS	Holmes, Sandra	703-648-6815
Hydrologic Hazards and Land Use	GS	Paulson, Richard	703-648-6851
Industrial Water Use	GS	Holmes, Sandra	703-648-6815
Irrigation Data, Farm and Ranch Lands	Census	Peterson, Dave	301-763-8560
Irrigation Water Use	GS	Holmes, Sandra	703-648-6815
Lakes, Biological Characteristics	ERL	Landers, Dixon	503-757-4427
Lakes, Chemical Characteristics	ERL	Landers, Dixon	503-757-4427
Lakes Sensitive to Acid Rain	ERL	Landers, Dixon	503-757-4427
Livestock Water Use	GS	Holmes, Sandra	703-648-6815
Marine Species, Spatial Distributions	NOS	LaPointe, Tom	301-443-0453
Marine Species, Temporal Distributions	NOS	Wolotira, Robert	301-443-0453
Mining Water Use	GS	Holmes, Sandra	703-648-6815
Navigable Waters, Spills, Potential Spills Data	CG	Robey, Mary	202-267-6670
Near Coastal Waters, Ecological Status (EMAP)	ORD	Latimer, Richard	401-782-3077
Nitrate Levels in Drinking Water Wells	OPP	Staff	800-426-4791
Ocean Data, Benthic Surveys	OW	King, Bob	202-260-7050
Ocean Data, Bioaccumulation Data	OW	King, Bob	202-260-7050
Ocean Data, Bioassays	OW	King, Bob	202-260-7050
Ocean Data, Effluent Data	OW	King, Bob	202-260-7050
Ocean Data, Environmental Conditions	OW	King, Bob	202-260-7050
Ocean Data, Fish Pathology	OW	King, Bob	202-260-7050
Ocean Data, Pollutant Discharges, Concentrations	OW	King, Bob	202-260-7050
Ocean Data, Quality of Receiving Water	OW	King, Bob	202-260-7050
Ocean Data, Sediment Chemistry	OW	King, Bob	202-260-7050
Ocean Data, Trawl Sampling	OW	King, Bob	202-260-7050
Oil Spills in U.S. Coastal Areas	CG	Staff	202-267-6993
Pesticide Concentrations in Drinking Wells	OPP	Staff	800-426-4791
Pollutant Discharge, Coastal Areas	NOS	Farrow, Daniel	301-443-0454
Pollution Abatement Activities in Manufacturing	Census	Shapiro, Janet	301-763-1755
Pollution Abatement, Operation Costs	Census	Shapiro, Janet	301-763-1755
Pollution Discharge Elimination Data (NPDES)	OTS	Sasnett, Samuel	202-260-1821
Pollution Sources	GS	Paulson, Richard	703-648-6851
Polution Abatement, Capital Expenditures	Census	Shapiro, Janet	301-763-1755
Public Water Supply, Water Use	GS	Holmes, Sandra	703-648-6815
Quality of Surface Water	GS	Miller, Timothy	703-648-6868

Subject	Source	Data Contact	Telephone
Radiation Levels in Drinking Water	NAREL	Goode, Paula	205-270-3433
Radiation Levels in Drinking Water	NAREL	Luster, Geraldine	205-270-3433
Radiation Levels in Surface Water	NAREL	Luster, Geraldine	205-270-3433
Radiation Levels in Surface Water	NAREL	Goode, Paula	205-270-3433
Radioactive Effluent, Nuclear Power Plants	BNL	Tichler, Joyce	516-282-3801
Radioactive Materials Releases, Nuclear Plants	BNL	Tichler, Joyce	516-282-3801
Radioactivity Associated with Drinking Water	NAREL	Petko, Charles	205-270-3400
Radioactivity Associated with Surface Water	NAREL	Petko, Charles	205-270-3400
Reservoir Contents Data from 100 Sites	GS	Ross, Thomas	703-648-6814
Rivers, Water Quality Data	GS	Ross, Thomas	703-648-6814
Rivers, Wild and Scenic Rivers, Number	NPS	Van Horne, Merle	202-343-3765
Seabird Colonies, Status, Commercial Harvests	NOS	Wolotira, Robert	301-443-0453
Sewerage Outlays by Government	Census	Wulf, Henry	301-763-7664
Shellfishing Waters, Bacterial Water Quality	NOS	Slaughter, Eric	301-443-8843
Spills and Potential Spills, Navigable Waters	CG	Robey, Mary	202-267-6670
Spills of Hazardous Substances in U.S. Waters	CG	Staff	202-267-6993
Streamflow Data from 190 Sites	GS	Ross, Thomas	703-648-6814
Streams, Data on Acidic Capacity	ERL	Landers, Dixon	503-757-4427
Surface Water, Acid Neutralizing Capacity	CERL	Stoddard, John	503-757-4427
Surface Water, Biological Characteristics	GS	Alexander, Richard	703-648-6869
Surface Water, Conditions and Trends	GS	Miller, Timothy	703-648-6868
Surface Water, Ecological Conditions (EMAP)	ORD	Latimer, Richard	401-782-3077
Surface Water, Large Scale Long-Term Trends	GS	Smith, Richard	703-648-6870
Surface Water, Physical Characteristics	GS	Smith, Richard	703-648-6870
Surface Water, Quantity, Quality	GS	Briggs, John	703-648-5624
Surface Water, Radiation Levels	NAREL	Luster, Geraldine	205-270-3433
Surface Water, Radiation Levels	NAREL	Goode, Paula	205-270-3433
Surface Water Unaffected by Human Activities	GS	Alexander, Richard	703-648-6869
Thermoelectric Power Generation, Water Use	GS	Holmes, Sandra	703-648-6815
Toxic Releases, Water Discharge	OTS	Sasnett, Samuel	202-260-1821
Utilities, Water, Revenues and Expenditures	Census	Wulf, Henry	301-763-7664
Water and Related Resources Data	SCS	George, Tommy	202-720-6267
Water Quality, Biological (STORET)	OW	Pandolfi, Thomas	202-260-7030
Water Quality, Chemical (STORET)	OW	Pandolfi, Thomas	202-260-7030
Water Quality, Global Data (Canada Contact)	CCIW	Allard, Martine	416-336-6441
Water Quality Monitoring	GS	Miller, Timothy	703-648-6868
Water Quality, Physical (STORET)	OW	Pandolfi, Thomas	202-260-7030
Water Resources, Summary Level Data	GS	Paulson, Richard	703-648-6851
Wetland Resources, Characteristics and Extent	FWS	Dahl, Thomas	813-893-3873
Wetland Resources, Characteristics and Extent	FWS	Wilen, Bill	703-358-2201

Subject	Source	Data Contact	Telephone
Wetland Resources, Status and Trends	FWS	Wilen, Bill	703-358-2201
Wetland Resources, Status and Trends	FWS	Dahl, Thomas	813-893-3873
Wetlands, Ecological Condition (EMAP)	ORD	Dixon, Thomas	202-260-5782
Wetlands, Vegetative Conditions	SCS	George, Tommy	202-720-6267
Wetlands, National Inventory Maps	FWS	Staff	800USAMAPS

Other Policy Relevant Environmental Data

Subject	Source	Data Contact	Telephone
Avian Population, Long-Term Trends	FWS	Peterjohn, Bruce	301-498-0330
Chemical Contaminants in Meat and Poultry	FSIS	Hubbert, William	202-205-0007
Chemical Hazard Exposure Information	OTS	Leitzke, John	202-260-3507
Contaminants, Toxic Elements in Fish and Birds	FWS	Steffeck, Donald	703-358-2148
Costs Recovered in Manufacturing	Census	Shapiro, Janet	301-763-1755
Endangered Species, Public Land Data	BLM	Ratliff, Michael	202-208-5717
Environmental Economic Statistics, National	CES	Shafer, Ronald	202-260-6966
Environmental Ecological Statistics, National	CES	Niemann, Brand	202-260-3726
Environmental Statistics, National	CES	Ross, Phil	202-260-2680
Fish and Birds, Contaminants, Toxic Elements	FWS	Steffeck, Donald	703-358-2148
Fish and Wildlife, Contaminant Threats	FWS	Steffeck, Donald	703-358-2148
Fish Hatcheries	FWS	Short, Olivia	703-358-1811
Food Quality, Global Data (U.S. Contact)	FDA	Burke, Jerry	202-245-1307
Hazardous Chemicals Exposure Data	OTS	Leitzke, John	202-260-3507
Health and Nutritional Status, U.S. Population	NCHS	Murphy, Robert	301-436-7068
Herbicide Use in Agricultural Crop Production	RFF	Gianessi, Leonard	202-328-5036
Hunting, Fishing and Wildlife Activities	FWS	Aiken, Richard	703-358-2156
Industrial Chemicals, Dietary Intake	FDA	Gunderson, Ellis	202-245-1152
Insect and Disease Conditions in Forest Lands	FS	Hofacker, Thomas	202-205-1600
Marine Mammals (Alaskan)	NIST	Wise, Stephen	301-975-3112
Meat and Poultry, Chemical Contaminants	FSIS	Hubbert, William	202-205-0007
Milk Samples, Radiation Levels	NAREL	Luster, Geraldine	205-270-3433
Milk Samples, Radiation Levels	NAREL	Goode, Paula	205-270-3433
Pesticide Concentrations in Water Wells	OPP	Staff	800-426-4791
Pesticide Data, Pesticide Info. Network (PIN)	OPP	Hoheisel, Constance	703-305-5455
Pesticide Residue on Raw and Processed Food	OPP	Reinert, Joseph	202-260-7557
Pesticides, Dietary Intake (Total Diet Study)	FDA	Gunderson, Ellis	202-245-1152
Radiation Levels in Milk Samples	NAREL	Luster, Geraldine	205-270-3433
Radiation Levels in Milk Samples	NAREL	Goode, Paula	205-270-3433
Radioactive Waste and Spent Fuel	ORNL	Klein, Jerry	615-574-6823
Radioactive Waste and Spent Fuel Inventory	OCRWM	Payton, M. L.	202-586-9140
Radioactivity Associated with Milk	NAREL	Petko, Charles	205-270-3400
Radionuclides, Dietary Intake	FDA	Gunderson, Ellis	202-245-1152
Radon Levels in Residences and Work Place	NAREL	Marcinowski, Frank	202-475-9615
Seabird Colonies, Status, Commercial Harvests	NOS	Wolotira, Robert	301-443-0453
Solar Radiation	NCDC	Staff	704-259-0682

Subject	Source	Data Contact	Telephone
Solar Radiation	NCDC	Staff	704-259-0682
Total Diet Study	FDA	Gunderson, Ellis	202-245-1152
Toxic Organic Compounds in the General Pop.	OTS	Remmers, Janet	202-260-1583
Toxic Waste Reduction Data	OTS	Sasnett, Samuel	202-260-1821
Waterfowl Breeding Population, Habitat Data	FWS	Blohm, Robert	703-358-1838
Waterfowl Production Areas	FWS	Short, Olivia	703-358-1811
Wildlife Populations, Public Lands Data	BLM	Ratliff, Michael	202-208-5717
Wildlife Refuges	FWS	Short, Olivia	703-358-1811

XIII. Energy Data Directory

Energy Data Sources

Uses of Energy Data

XIII. Energy Data Directory

Energy Data Directory

The Energy Data Directory guides you to the most important energy-business data managers. It consists of two sections: Energy Data Sources and Uses of Energy Data. The Directory covers **coal, crude oil natural gas, nuclear power, utilities, steam, refineries, vehicles, pipelines and other major topics in the energy area.**

The primary source for energy data is the National Energy Information Center (NEIC). NEIC can provide you with energy-relevant statistical and analytical data, information and referral assistance. You can request the free-of-charge Energy Information Administration press releases to stay on top of emerging issues.

For general information on the energy issues listed below, contact the following data experts at: **(202) 586-8800.**

Coal: Dorothy Karsteter or Bill Horvath

Electric and Nuclear Power: Thomas Welch or William Jeffers

Natural Gas: Paula Altman or Jonathan Cogan

Petroleum: Leola Withrow or Trisha Christian

Renewable Energy Resources and Conservation: Marion King or Karen Freedman

Section I. Energy Data Sources

Major Categories

Coal and Coke Products
Crude Oil
Electric Plants and Electric Utilities
Natural Gas
Oil and Petroleum Products
Other Energy Sources

Organizational Index and Acronyms

EIA	Energy Information Administration, U.S. Department of Energy
FE	Fossil Energy Administration, U.S. Department of Energy
FERC	Federal Energy Regulatory Commission, U.S. Department of Energy
FPC	Federal Power Commission, U.S. Department of Energy
NEIC	National Energy Information Center, U.S. Department of Energy

Energy Data Sources

Subject	Source	Data Contact	Telephone

Coal and Coke

Coal, Consumption, Stocks, Prices and Sales	EIA	Paull, Mary	202-254-5379
Coal, Employment and Projected Production	EIA	Freme, Fred	202-254-5379
Coal, End of Quarter Stocks	EIA	Paull, Mary	202-254-5379
Coal, Imports, Quantity, Quality and Cost	EIA	McClevey, Ken	202-254-5655
Coal, Origin, Distribution and Sales	EIA	Paull, Mary	202-254-5379
Coal, Production Quantity and Value	EIA	Freme, Fred	202-254-5379
Coal, Productive Capacity	EIA	Freme, Fred	202-254-5379
Coal, Recoverable Reserves	EIA	Freme, Fred	202-254-5379
Coal, Sales and Stocks of Coke	EIA	Paull, Mary	202-254-5379
Coal, Types of Mining Operations	EIA	Freme, Fred	202-254-5379
Coal, Use by U.S. Manufacturing Plants, Ranked	EIA	Paull, Mary	202-254-5379
Coke, Production, Transfers, and Consumption	EIA	Paull, Mary	202-254-5379

Crude Oil

Crude Oil, Alaskan, Stocks in Transit by Pipeline	EIA	Gray, Christine	202-586-8995
Crude Oil, Alaskan, Stocks in Transit by Water	EIA	Gray, Christine	202-586-8995
Crude Oil, Consumption by Pipelines	EIA	Gray, Christine	202-586-8995
Crude Oil, End-of-Month Stocks	EIA	Gray, Christine	202-586-8995
Crude Oil, Foreign, Amounts Purchased	EIA	Scott, Elizabeth	202-586-1258
Crude Oil, Foreign, Purchase Price Data	EIA	Scott, Elizabeth	202-586-1258
Crude Oil, Foreign, Transactions by Country and Type	EIA	Scott, Elizabeth	202-586-1258
Crude Oil, Imports into the U.S.	EIA	Myers, Stacey	202-586-5130
Crude Oil, Proved Reserves and Production	EIA	Wood, John	214-767-2200
Crude Oil Purchased by Refiners, Cost Data	EIA	Scott, Elizabeth	202-586-1258
Oil in Transit and Stocks at Sea	EIA	Myers, Stacey	202-586-5130
Oil Recovery Projects, Size and Type of Test	FE	Allison, Edith	918-336-4390

Part Two: The DATAPHONER Energy Data Sources

Subject	Source	Data Contact	Telephone

Electric Plants and Electric Utilities

Subject	Source	Data Contact	Telephone
Electric Generating Plants, Data by Type of Plant	EIA	Russell, Charlene	202-254-5437
Electric Generating Plants, Data on Power Plant Sites	EIA	Bess, Elsie	202-254-5637
Electric Generating Plants, Fuel Consumption	EIA	Johnson, Melvin	202-254-5665
Electric Generating Plants, Fuel Stocks	EIA	Johnson, Melvin	202-254-5665
Electric Plants, Quantity, Quality of Fuel Received	FERC	McClevey, Ken	202-254-5655
Electric Plants, Type of Purchases	FERC	McClevey, Ken	202-254-5655
Electric Power Plants, Steam, Air Quality Data	EIA	Breuel, A.	202-254-5628
Electric Utilities, Electric Sales and Purchases	EIA	Harris, Charlene	202-254-5437
Electric Utilities, Major, Operating Expense Data	FERC	Secquety, Roger	202-254-5556
Electric Utilities, Public-Owned, Financial Data	EIA	Smith, Susan	202-254-5668
Electrical Import and Export Data	FE	Mintz, Steven	202-586-9506
Electricity Sales at the State Electric Utility Level	EIA	Calopedis, Stephen	202-254-5661
Steam Generating Units, Coal-Fired	EIA	Bess, Elsie	202-254-5637

Natural Gas

Subject	Source	Data Contact	Telephone
Natural Gas Imports and Exports, Volume, Value	FPC	Dillard, Fay	202-586-6181
Natural Gas, Injections, Withdrawals	EIA	Maupin, Ellis	202-586-6178
Natural Gas Liquids, Imports into the U.S.	EIA	Myers, Stacey	202-586-5130
Natural Gas Liquids, Reserves, Production Data	EIA	Wood, John	214-767-2200
Natural Gas Liquids, Stocks, Receipts	EIA	Hinton, Dave	202-586-2990
Natural Gas Liquids, Stocks, Receipts	EIA	Hinton, Dave	202-586-2990
Natural Gas, Location and Capacity of Reservoirs	EIA	Maupin, Ellis	202-586-6178
Natural Gas, Number of Producing Gas Wells	EIA	Dunston, Donna	202-586-6135
Natural Gas, Origins of Natural Gas Received	EIA	Wood, John	214-767-2200
Natural Gas, Origins of Natural Gas Supplies	EIA	Natof, Margo	202-586-6303
Natural Gas Pipelines, Financial Data	FERC	Mack, Juanita	202-586-6169
Natural Gas Pipelines Rates	FERC	Keeling, James	202-586-6107
Natural Gas Pipelines, Revenue and Expense Data	FERC	Keeling, James	202-586-6107
Natural Gas Pipelines, Supplies and Production	FERC	Keeling, James	202-586-6107
Natural Gas, Proved Reserves and Production	EIA	Wood, John	214-767-2200
Natural Gas Sales to Customers	FERC	Keeling, James	202-586-6107
Natural Gas, State Disposition	EIA	Natof, Margo	202-586-6303
Natural Gas, Underground Storage	EIA	Maupin, Ellis	202-586-6178
Natural Gas, Value of Production	EIA	Dunston, Donna	202-586-6135
Natural Gas, Working and Base Gas in Reservoirs	EIA	Maupin, Ellis	202-586-6178

Subject	Source	Data Contact	Telephone

Oil and Petroleum Products

Fuel Oil, Distillate and Residual, Annual Sales Data	EIA	Lippert, Alice	202-586-9600
Fuel Oil Sold by Resellers and Retailers	EIA	Riner, Charles	202-586-6610
Gasoline, Data on State Sales	EIA	Riner, Charles	202-586-6610
Gasoline, Volume and Prices	EIA	Riner, Charles	202-586-6610
Kerosene, Annual Sales Data	EIA	Lippert, Alice	202-586-9600
Petroleum Products, Data on Sales and Prices	EIA	Riner, Charles	202-586-6610
Petroleum Products, Imports into the U.S	EIA	Graham, Claudette	202-586-9649
Petroleum Products, Refined Stocks, Production Data	EIA	Masterson, Nancy	202-586-8393
Petroleum Refineries, Blending Plants, Capacity Data	EIA	Masterson, Nancy	202-586-8393

Other Energy Sources

Energy Consumption in Commercial Buildings	EIA	Johnson, Martha	202-586-1135
Energy Consumption in the Residential Sector	EIA	Thompson, W.	202-586-1119
Energy Consumption In the Manufacturing Sector	EIA	Preston, John	202-586-1128
Energy Expenditures in Commercial Buildings	EIA	Johnson, Martha	202-586-1135
Energy Expenditures in the Residential Sector	EIA	Thompson, W.	202-586-1119
Nuclear Reactor Construction Cost	EIA	Smith, Luther	202-254-5565
Nuclear Reactors, Commercial Operation Data	EIA	Smith, Luther	202-254-5565

Uses of Energy Data | *XIII. Energy Data Directory*

Section II. Uses of Energy Data

Description and Source of Energy Data	Uses of Energy Data
Coal, consumption, stocks, prices and quality by rank at U.S. manufacturing plants. Coal Consumption Report, Quarterly, EIA -3 **Data Manager: Mary K. Paull** **Telephone: (202) 254-5379**	Used for coal demand analysis and forecasting of the demand and prices for coal.
Coal, origin, distribution, sales and end of quarter stocks. Coal Distribution Report, EIA -6 **Data Manager: Mary K. Paull** **Telephone: (202) 254-5379**	Used to produce statistical reports, publications and analysis.
Coal, types of mining operations, recoverable reserves, production quantity and value, productive capacity, employment and projected production. Coal Production Report, EIA -7A **Data Manager: Fred Freme** **Telephone: (202) 254-5367**	Used to provide information on current and prospective coal production, capacity, prices, reserves, and labor productivity.
Coal Imports, quantity, quality and cost of coal imported by electric utility plants. Quarterly Coal Imports by Electric Utilities Into the United States, EIA -868 **Data Manager: Ken McClevey** **Telephone: (202) 254-5655**	Used to prepare quarterly and annual summaries of coal imports, including data and analysis on prices, sources of exports, transportation methods and costs and displacement of U.S. coal by imports.

213

Description and Source of Energy Data	Uses of Energy Data
Coke, production, transfers, consumption, sales and stocks of coal and coke. Coke Plant Report, Quarterly, EIA -5 **Data Manager: Mary K. Paull** **Telephone: (202) 254-5379**	Used to produce statistical reports, publications and analysis.
Crude Oil, cost of crude oil purchased by refiners. Refiners' Monthly Cost Report, EIA -14 **Data Manager: Elizabeth Scott** **Telephone: (202) 586-1258**	Used to measure and track crude oil purchased by refiners.
Crude Oil, proved reserves and production of crude oil. Annual Survey of Domestic Oil and Gas Reserves, EIA -23 **Data Manager: John Wood** **Telephone: (214) 767-2200**	Used to estimate national and regional data on the reserves of crude oil and to facilitate national energy policy decisions.
Crude Oil, data on end-of-month stocks of crude oil, consumption of crude oil during the month by pipelines and on leases, stocks of Alaskan crude oil in transit by water and pipelines. Monthly Crude Oil Report, EIA -813 **Data Manager: Christine Gray** **Telephone: (202) 586-8995**	Used to monitor petroleum supply and demand.
Crude Oil, Foreign, data on crude oil transactions, country crude code, crude type, gravity, data on loading/landing, port of destination, vessel, volume purchased, purchase price and other related data. Monthly Foreign Crude Oil Acquisition Report, EIA -856 **Data Manager:** Elizabeth Scott **Telephone: (202) 586-1258**	Used to track the costs of foreign crude oil acquired for importation into the U.S. and its territories and possessions; to calculate price indices; to analyze consumption, production, and prices of fuels worldwide; and in modeling and forecasting.

Uses of Energy Data   *XIII. Energy Data Directory*

Description and Source of Energy Data	Uses of Energy Data
Crude Oil Imports, data on imports of crude oil. Monthly Imports Report, EIA -814 **Data Manager: Claudette Graham** **Telephone: (202) 586-9649**	Used to monitor petroleum supply and demand.
Electric Generating Plants, net generation, fuel consumption, and end-of-month fuel stocks for all electric generating plants. Monthly Power Plant Report, EIA -759 **Data Manager: Melvin Johnson** **Telephone: (202) 254-5665**	Used in forecasting models.
Electric Generating Plants, data on power plant sites, generators, and milestones for coal-fired steam generating units. Annual Electric Generator Report, EIA -860 **Data Manager: Elsie Bess** **Telephone: (202) 254-5637**	Used to monitor the status of electric generating plants and associated equipment in commercial service as well as those scheduled to be in service within 10 years of filing of the reports.
Electric Plants, type of purchase, expiration data of contract, fuel type, coal origin data, quantity and quality of fuel received. Monthly Report of Cost and Quality of Fuels for Electric Plants, FERC -423 **Data Manager: Ken McClevey** **Telephone: (202) 254-5655**	Used in economic studies to determine the justification for increasing electric rates, environmental studies, fuel emergencies, and policy decisions.
Electric Plants, Steam, air and water quality data from steam-electric power plants. Steam-Electric Plant Operation and Design Report, EIA -767 **Data Manager: Al Breuel** **Telephone: (202) 254-5628**	Used to evaluate fuel use in rate proceedings; to develop, assess, reform, and enforce regulations; to assess the impact of pollution abatement and control expenditures on the GDP; and to assess the effect of environmental regulations on the generation of electric power.

Description and Source of Energy Data	Uses of Energy Data
Electric Plants, Steam, air and water quality data from steam-electric power plants. Steam-Electric Plant Operation and Design Report, EIA -767 **Data Manager: Al Breuel** **Telephone: (202) 254-5628**	Used to evaluate fuel use in rate proceedings; to develop, assess, reform, and enforce regulations; to assess the impact of pollution abatement and control expenditures on the GDP; and to assess the effect of environmental regulations on the generation of electric power.
Electric Utilities, financial statistics, electric sales and purchases, generating plant data by type of plant and transmission line data. Annual Report of Public Electric Utilities, EIA -412 **Data Manager: Charlene Harris-Russell** **Telephone: (202) 254-5437**	Used to provide accounting, financial, and operating data of publicly owned electric utilities.
Electric Utilities, electricity sales and associated revenue to ultimate consumers by class of service at the state electric utility level and related financial data. Monthly Electric Utility Sales and Revenue Report with State Distributions, EIA -826 **Data Manager: Stephen Calopedis** **Telephone: (202) 254-5661**	Used to estimate sales, and average revenue per kilowatthour sold at the national, regional and state level. The data is also used in the compilation of the Gross Domestic Product.
Electric Utilities, financial statistics, electric sales and revenue data for public and investor-owned electric utilities. Annual Electric Utility Report, EIA -861 **Data Manager: Susan Smith** **Telephone: (202) 254-5668**	Used to answer a wide range of questions on the generation, transmission, distribution and sale of electric energy primarily for use by the public.

Uses of Energy Data   *XIII. Energy Data Directory*

Description and Source of Energy Data	Uses of Energy Data
Electric Utilities, Major, financial data, operating revenues, electric maintenance expenses, and generating plant statistics. Annual Report of Major Electric Utilities, Licensees and Others, FERC-1 **Data Manager: Roger Secquety** **Telephone: (202) 254-5556**	Used for formal investigation of electric rates, rate levels, and financial audits.
Electrical Imports and Exports, electrical import and export data. Annual Report of International Electrical Export/Import Data, FE -781R **Data Manager: Steven Mintz** **Telephone: (202) 586-9506**	Used to track electricity imported into the U.S. and to furnish decision makers data on which to base trade policy.
Energy Consumption, Residential, data on consumption of, and expenditure for, energy in the residential sector of the economy including housing and demographic characteristics. Residential Energy Consumption Survey, EIA -457A/G **Data Manager: Wendel Thompson** **Telephone: (202) 586-1119**	Used for analyzing and forecasting residential energy consumption.
Energy Use, Commercial Buildings, consumption of, and expenditures for, energy in the commercial building sector of the economy. Commercial Buildings Energy Consumption Survey, EIA -871A/H **Data Manager: Martha Johnson** **Telephone: (202) 586-1135**	Used in energy models and as the basis for statistical and analytical reports on the characteristics of commercial buildings use of energy.

Part Two: The DATAPHONER							Uses of Energy Data

Description and Source of Energy Data	Uses of Energy Data
Energy Use, Manufacturing, manufacturing consumption of energy for fuel and nonfuel purposes, fuel-switching capabilities, and energy prices, on-site electricity generation and purchases of electricity from non-utilities. Manufacturing Energy Consumption Survey, EIA -846A/D **Data Manager: John L. Preston** **Telephone: (202) 586-1128**	Used to monitor the energy consumption, energy usage patterns, and fuel-switching capabilities of the manufacturing sector of the U.S. economy.
Fuel Oil, Distillate and Residual, annual sales data. Annual Fuel Oil and Kerosene Sales Report, EIA -821 **Data Manager: Alice Lippert** **Telephone: (202) 586-9600**	Used to determine current and projected fuel oil needs on national, regional and state levels.
Gas Storage, Underground, working and base gas in reservoirs, injections, withdrawals and location and capacity of reservoirs. Underground Gas Storage Report, EIA -191 **Data Manager: Ellis Maupin** **Telephone: (202) 586-6178**	Used to assess the supplies of natural gas in storage fields in regions of the U.S. and to identify the location of supplies.
Kerosene, annual sales data. Annual Fuel Oil and Kerosene Sales Report, EIA -821 **Data Manager: Alice Lippert** **Telephone: (202) 586-9600**	Used to determine current and projected fuel oil needs on national, regional and state levels.
Natural Gas, proved reserves and production of natural gas and natural gas liquids. Annual Survey of Domestic Oil and Gas Reserves, EIA -23 **Data Manager: John Wood** **Telephone: (214) 767-2200**	Used to estimate national and regional data on the reserves of natural gas and natural gas liquids and to facilitate national energy policy decisions.

Uses of Energy Data   *XIII. Energy Data Directory*

Description and Source of Energy Data	Uses of Energy Data
Natural Gas, origins of natural gas supplies and the disposition of natural gas on a state basis. Annual Report of Natural and Supplemental Gas Supply and Disposition, EIA -176 **Data Manager: Margo Natof** **Telephone: (202) 586-6303**	Used to estimate the consumption of natural gas by major end-use category, demand, and prices by state for analysis and publication.
Natural Gas, natural gas production, value of natural gas, and the number of producing gas wells. Annual Quantity and Value of Natural Gas Report, EIA -627 **Data Manager: Donna Dunston** **Telephone: (202) 586-6135**	Used to estimate natural gas production, the value of natural gas, and the number of producing gas wells.
Natural Gas Imports and Exports, data on transporters, U.S. point of entry, foreign buyer or seller, docket number, and volume and dollar amount of natural gas exports and imports. Annual Report for Importers and Exporters of Natural Gas, FPC-14 **Data Manager: Fay Dillard** **Telephone: (202) 586-6181**	Used to help monitor and regulate natural gas imports into, and exports from, the U.S.
Natural Gas Liquids, origins of natural gas received and natural gas plant liquids produced. Annual Report of the Origin of Natural Gas Liquids Production, EIA -64A **Data Manager: John Wood** **Telephone: (214) 767-2200**	Used to estimate natural gas plant liquids production and reserves by state and region.
Natural Gas Liquids, beginning stocks, receipts, and production and input, shipments, fuel use and losses and endings stocks of natural gas liquids. Monthly Natural Gas Liquids Report, EIA -816 **Data Manager: Dave Hinton** **Telephone: (202) 586-2990**	Used to report aggregate statistics on, and conduct analyses of, the operation of U.S. natural gas processing plants and fractionators.

219

Part Two: The DATAPHONER   Uses of Energy Data

Description and Source of Energy Data	Uses of Energy Data
Natural Gas Pipelines, financial data and operating expenses. Annual Report for Major Natural Gas Companies, FERC -2 **Data Manager: Juanita Mack** **Telephone: (202) 586-6169**	Used for gas pipeline review and rate-setting.
Natural Gas Pipelines, data on revenues, expenses, and gas volume, end-of-month sales of natural gas to customers, income, operation and maintenance expenses, rates, and gas supplies and production. Natural Gas Pipeline Company Monthly Statement, FERC -11 **Data Manager: James Keeling** **Telephone: (202) 586-6107**	Used to measure the financial status of the regulated pipelines as a group.
Nuclear Reactor Construction, cost data, date of first fuel loading, date unit is scheduled for commercial operation. Semiannual Report on Status of Reactor Construction, EIA -254 **Data Manager: Luther Smith** **Telephone: (202) 254-5565**	Used to provide data on nuclear units planned or under construction by electric utilities.
Oil Imports, imports of crude oil, natural gas liquids and petroleum products into the U.S., oil in transit, and stocks at sea. International Energy Agency Imports / Stocks-at-Sea Report, EIA -818 **Data Manager: Stacey Myers** **Telephone: (202) 586-5130**	Used in estimating actual and projected oil imports. During petroleum supply emergencies, the data are used to help re-allocate petroleum among the International Energy Agency member nations.

Uses of Energy Data   *XIII. Energy Data Directory*

Description and Source of Energy Data	Uses of Energy Data
Oil Recovery Projects, size and type of test, reservoir description, injection and produced fluid quantities. Enhanced Oil Recovery Annual Report, FE -748 **Data Manager: Edith Allison** **Telephone: (918) 336-4390**	Used for tracking the progress of enhanced oil recovery projects.
Petroleum Products, data on sales, prices and volumes of selected petroleum products of refiners and gas plant operators. Refiners / Gas Plant Operators' Monthly Petroleum Product Sales Report, EIA -782A **Data Manager: Charles Riner** **Telephone: (202) 586-6610**	Used as a basis for analysis and to make projections related to energy supplies, demand and prices.
Petroleum Products, data on state sales volume and prices for motor gasoline, No. 2 distillate and residual fuel oil sold by resellers and retailers. Resellers' / Retailers' Monthly Petroleum Product Sales Report, EIA -782B **Data Manager: Charles Riner** **Telephone: (202) 586-6610**	Used as a basis for analysis and to make projections related to energy supplies, demand and prices.
Petroleum Products, sales of selected petroleum products into a state. Monthly Report of Petroleum Products Sold into a State for Consumption, EIA -782C **Data Manager: Kenneth Platto** **Telephone: (202) 586-6364**	Used to perform analysis and make projections related to energy supply and demand.
Petroleum Products, Imports, data on imports of petroleum products. Monthly Imports Report, EIA -814 **Data Manager: Claudette Graham** **Telephone: (202) 586-9649**	Used to monitor petroleum supply and demand.

Description and Source of Energy Data	Uses of Energy Data
Petroleum Products, Refined, beginning stocks, receipts, and production and input, shipments, fuel use and losses and ending stocks of refined petroleum products. Monthly Refinery Report, EIA -810 Data Manager: Nancy Masterson Telephone: (202) 586-8393	Used to analyze the balance between the supply and disposition of refined products produced by all refineries and blending plants.
Petroleum Refineries and Blending Plants, data on current and projected capacities of the facilities of all U.S. petroleum refineries and blending plants. Annual Refinery Report, EIA -820 Data Manager: Nancy Masterson Telephone: (202) 586-8393	Used to generate aggregate statistics on, and to conduct analysis of, the operation of U.S. petroleum refineries and blending plants.
Uranium Industry, data on uranium exploration and development, reserves, ore and concentrate production, marketing and inventory data, shipments for enrichment, requirements, and financial data on the uranium industry in the U.S. Uranium Industry Annual Survey, EIA -858 Data Manager: Luther Smith Telephone: (202) 254-5565	Used to monitor the viability of the domestic uranium mining and milling industry pursuant to the Nuclear Regulatory Commission Authorization Act of 1983. The data are used extensively by the public and private sectors to analyze trends in the uranium industry and to assess its current status.
Vehicle Use, Households, data on the number and types of vehicles per household, mileage per vehicle, vehicle characteristics and fuel types used. Residential Transportation Energy Consumption Survey, EIA -876A/E Data Manager: Ronald Lambrecht Telephone: (202) 586-4962	Used to estimate vehicle fuel consumption, expenditures, and fuel efficiency.

XIV. Index of Data Experts

Index of Data Experts

Organizational Index and Acronyms

AEERL	Air and Energy Engineering Research Laboratory, U.S. Environmental Protection Agency
ANL	Argonne National Laboratory, U.S. Department of Energy
AREAL	Atmospheric Research and Exposure Assessment Laboratory, U.S. Environmental Protection Agency
ATSDR	Agency for Toxic Substances and Disease Registry
BEA	Bureau of Economic Analysis, U.S. Department of Commerce
BLM	Bureau of Land Management, U.S. Department of the Interior
BLS	Bureau of Labor Statistics, U.S. Department of Labor
BNL	Brookhaven National Laboratory
BOM	Bureau of Mines, U.S. Department of the Interior
CCIW	Canada Centre for Inland Water
CENSUS	Bureau of the Census, U.S. Department of Commerce
CERL	Corvallis Environmental Research Laboratory, U.S. Environmental Protection Agency
CES	Center for Environmental Statistics, U.S. Environmental Protection Agency
CG	Coast Guard, U.S. Department of Transportation
CTI	Center for Transportation Information, U.S. Department of Transportation
DHHS	U.S. Department of Health and Human Services
EBRI	Employee Benefit Research Institute
EIA	Energy Information Administration, U.S. Department of Energy
ERL	Environmental Research Laboratory, U.S. Environmental Protection Agency
ERS	Economic Research Service, U.S. Department of Agriculture
FDA	Food and Drug Administration, U.S. Department of Health and Human Services
FE	Fossil Energy Administration, U.S. Department of Energy
FERC	Federal Energy Regulatory Commission, U.S. Department of Energy
FHA	Federal Highway Administration, U.S. Department of Transportation
FPC	Federal Power Commission, U.S. Department of Energy
FRS	Federal Reserve System
FS	U.S. Forest Service, U.S. Department of Agriculture
FSIS	Food Safety and Inspection Service, U.S. Department of Agriculture
FWS	U.S. Fish and Wildlife Service, U.S. Department of the Interior
GS	Geological Survey, U.S. Department of the Interior
HA	Hewitt Associates
HCBS	Health Care Benefits Survey
HCFA	Health Care Financing Administration, U.S. Department of Health and Human Services
HIAA	Health Insurance Association of America

Part Two: The DATAPHONER

NAREL	National Air and Radiation Environmental Laboratory, U.S. Environmental Protection Agency
NASS	National Agricultural Statistics Service, U.S. Department of Agriculture
NCDC	National Climate Data Center, NOAA, U.S. Department of Commerce
NCHS	National Center for Health Statistics, U.S. Department of Health and Human Services
NEIC	National Energy Information Center, U.S. Department of Energy
NFIB	National Federation of Independent Businesses
NIAAA	National Institute of Alcohol Abuse and Alcoholism
NMFS	National Marine Fisheries Service, NOAA, U.S. Department of Commerce
NOAA	National Oceanic and Atmospheric Administration, U.S. Department of Commerce
NOS	National Ocean Service, NOAA, U.S. Department of Commerce
NPS	National Park Service, U.S. Department of the Interior
OAQPS	Office of Air Quality Planning and Standards, U.S. Environmental Protection Agency
OBA	Office of Business Analysis, U.S. Department of Commerce
OCRWM	Office of Civilian Radioactive Waste Management, U.S. Department of Energy
OERR	Office of Emergency and Remedial Response (Superfund), U.S. Environmental Protection Agency
OGDW	Office of Groundwater and Drinking Water, U.S. Environmental Protection Agency
OMS	Office of Mobile Sources, U.S. Environmental Protection Agency
OPP	Office of Pesticide Programs, U.S. Environmental Protection Agency
ORD	Office of Research and Development, U.S. Environmental Protection Agency
ORNL	Oak Ridge National Laboratory, U.S. Department of Energy
OSW	Office of Solid Waste, U.S. Environmental Protection Agency
OTS	Office of Toxic Substances, U.S. Environmental Protection Agency
OW	Office of Water, U.S. Environmental Protection Agency
PSRF	Profit Sharing Research Foundation
PWBA	Pension and Welfare Benefits Administration, U.S. Department of Labor
RFF	Resources for the Future
SBA	Small Business Administration
SCS	Soil Conservation Service, U.S. Department of Agriculture
WB-DOL	Women's Bureau, U.S. Department of Labor
WLDF	Women's Legal Defense Fund

We are using the following three acronyms to reference the main directories of this book:

BUS	Business Data Directory
ENE	Energy Data Directory
ENV	Environmental Data Directory

Index of Data Experts

Name	Data Directory	Subject	Agency	Telephone
Abney, Willie	BUS	Proprietors' Income	BEA	202-523-0811
Ackerman, Karen	BUS	Agricultural Export Programs	ERS	202-219-0820
Adams, Don	BUS	Data Mgr., Foreign Trade	Census	301-763-5342
Adams, Donald	BUS	Gold	FRS	202-452-2364
Adkins, Robert	BUS	Fuel Price Indexes	BLS	202-272-5177
Ahearn, Mary	BUS	Farm Household Income	ERS	202-219-0807
Aiken, Richard	ENV	Hunting, Fishing and Wildlife	FWS	703-358-2156
Alexander, Richard	ENV	Water Quality and Usage	GS	703-648-6869
Allard, Martine	ENV	Water Quality Monitoring	CCIW	416-336-6441
Allen, Ed	BUS	Wheat, Prices and Economic Data	ERS	202-219-0841
Allison, Edith	ENE	Oil Recovery Projects	FE	918-336-4390
Alterman, William	BUS	Data Mgr., International Prices	BLS	202-272-5020
Amel, Dean	BUS	Banking Structure	FRS	202-452-2911
Anderson, John	BUS	Personal Income and Related Data	BLS	202-501-6463
Andreassen, Art	BUS	Supplemental Workers	BLS	202-272-5326
Ardolini, Charles	BUS	Productivity Trends	BLS	202-523-9244
Armknecht, Paul	BUS	Data Mgr., Consumer Prices	BLS	202-272-5164
August, James	BUS	Consumer Credit, Analysis	FRS	202-452-3741
Ayres, Mary Ellen	BUS	Computerized Labor Data	BLS	202-523-7827
Bach, Christopher	BUS	Data Mgr., Balance of Payments	BEA	202-523-0620
Bachu, Amara	BUS	Population, Births and Fertility	Census	301-763-5303
Bailey, Linda	BUS	World Agricultural Data	ERS	202-219-1286
Bailey, Wallace	BUS	Employment, Subnational Data	BEA	202-254-6635
Baldwin, Godfrey	BUS	International and Foreign Trade	Census	301-763-4022
Baldwin, Steve	BUS	Business Inventories	BEA	202-523-0784
Banister, Judith	BUS	Foreign Trade, China, CPR	Census	301-763-4012
Barker, Betty	BUS	Foreign Direct Investment	BEA	202-523-0659
Barna, John	BUS	Foreign Trade Classifications	Census	301-763-7766
Barrett, Fred	BUS	Data Mgr., State Agricultural Data	NASS	202-720-3638
Bauman, Alvin	BUS	Data Mgr., Labor-Mgmt. Relations	BLS	202-523-1143
Baxter, Leila	BUS	Map Orders, 1990 Census	Census	812-288-3192
Baxter, Tim	BUS	Agricultural Credit and Financial Markets	ERS	202-219-0706
Beale, Calvin	BUS	Population	ERS	202-219-0535
Becker, Gerald	BUS	International and Foreign Trade Data	Census	301-763-7126
Beckman, Barry	BUS	Business Cycle Methodology	BEA	202-523-0800

Part Two: The DATAPHONER

Name	Data Directory	Subject	Agency	Telephone
Bell, John	BUS	Profit Sharing Plans	PSRF	312-372-3416
Beller, Dan	BUS	ERISA	PWBA	202-523-9505
Berman, Patricia	BUS	Decennial Census Program Design	Census	301-763-7094
Berman, Steve	BUS	Residential Construction, New	Census	301-763-7842
Bernat, Andrew	BUS	Rural Development Data	ERS	202-219-0540
Bernhardt, Malcolm	BUS	Current Industrial Reports (CIRs)	Census	301-763-2518
Bernstein, Robert	BUS	Statistical Briefs	Census	301-763-1584
Bertelsen, Diane	BUS	Fruits and Tree Nuts	ERS	202-219-0884
Bess, Elsie	ENE	Electric Plants and Utilities	EIA	202-254-5637
Bethea, Martha	BUS	C&CA Reserve Requirement	FRS	202-452-3181
Bettge, Paul	BUS	Balance Sheet, Interest on FR Notes	FRS	202-452-3174
Bezirganian, Steve	BUS	Foreign Direct Investment	BEA	202-523-0641
Biederman, Edna	BUS	Labor Force Demographics	BLS	202-523-1002
Bird, Dick	ENV	Land Use	BLM	202-653-8864
Blackledge, John	BUS	Horticulture Statistics	Census	301-763-8560
Blake, Pamela	BUS	Health Care Cost	HCBS	609-520-2289
Blaylock, James	BUS	Food Demand and Expenditures	ERS	202-219-0862
Blohm, Robert	ENV	Waterfowl Population	FWS	703-358-1838
Blostin, Allan	BUS	Health Insurance Benefits	BLS	202-523-8791
Blue, Eunice	BUS	GNP Computer Tapes	BEA	202-523-0804
Blyweiss, Hal	BUS	Shippers' Export Information	Census	301-763-5310
Boertlein, Celia	BUS	Commuting, Place of Work	Census	301-763-3850
Bolyard, Joan	BUS	Travel, International	BEA	202-523-0609
Bomkamp, James	BUS	Foreign Direct Investment	BEA	202-523-0559
Bonds, Belinda	BUS	Input-Output Tables	BEA	202-523-0843
Bones, James	ENV	Forest Resources	FS	202-205-1343
Bontekoe, Eldert	ENV	Pollutants from Motor Vehicles	OMS	313-668-4200
Bostic, William	BUS	Truck Inventory and Use	Census	301-763-2735
Bowers, Douglas	BUS	Agricultural History	ERS	202-219-0787
Bowie, Chester	BUS	SIPP Income Survey	Census	301-763-2764
Bowman, Charles	BUS	Economic Growth	BLS	202-272-5383
Bowman, Sharon	BUS	Electronic Funds Transfers	FRS	202-452-3867
Braden, Brad	BUS	Employee Compensation Cost	BLS	202-523-1165
Bradley, Kathleen	ENV	Transportation Statistics	CTI	617-494-2614
Branscome, Jim	BUS	Data Mgr., Consumer Price Index	BLS	202-272-2322
Braun, Steven	BUS	Price and Inflation Developments	FRS	202-452-3800
Brayton, Flint	BUS	Quarterly Econometric Model	FRS	202-452-2670
Brett, Gayle	BUS	Transfer of Funds	FRS	202-452-2934
Breuel, Al	ENE	Electric Plants and Utilities	EIA	202-254-5628
Brewster, Jim	BUS	Floriculture	NASS	202-219-7688

Index of Data Experts

Name	Data Directory	Subject	Agency	Telephone
Briggs, John	BUS	Groundwater Quality and Usage	GS	703-648-5624
Broome, Fred	BUS	Computer Mapping	Census	301-763-3973
Brown, Anita	BUS	Automated Foreign Trade Data	Census	301-763-7700
Brown, Robert	BUS	Metropolitan Area Personal Income	BEA	202-254-6632
Brown, Sharon	BUS	Data Mgr., Local Unemployment	BLS	202-523-1038
Buche, John	BUS	Farm Prices, Parity Received	NASS	202-720-5446
Buckley, John	BUS	Area Wage Survey	BLS	202-523-1763
Buckner, Dan	BUS	Dairy Product Statistics	NASS	202-720-4448
Budge, Arvin	BUS	Dry Edible Beans Data	NASS	202-720-4285
Burke, James	BUS	BHCs, Analysis of Mergers/Acquisitions	FRS	202-452-2612
Burke, Jerry	ENV	Food Quality Monitoring	FDA	202-245-1307
Bush, Joseph	BUS	Wages by Industry, State and Area	BLS	202-523-1158
Buso, Michael	BUS	BLS Data Diskettes and Tapes	BLS	202-523-1158
Buzzanell, Peter	BUS	Sugar, Prices and Economic Data	ERS	202-219-0888
Cagle, Maury	BUS	Data Mgr., Public Information	Census	301-763-4040
Calopedis, Stephen	ENE	Electricity Sales and Revenue	EIA	202-254-5661
Campell, Carmen	BUS	Public-Use Microdata Samples	Census	301-763-2005
Cannon, John	BUS	Travel Surveys	Census	301-763-5468
Capdevielle, Patricia	BUS	Foreign Compensation Cost	BLS	202-523-9292
Capehart, Tom	BUS	Tobacco, Prices and Economic Data	ERS	202-219-0890
Carbaugh, Larry	BUS	State Data Center Program	Census	301-763-1580
Carlson, Norma	BUS	Work Injury Reports Survey	BLS	202-501-7570
Carlson, Robert	BUS	Data Mgr., Establishment Data	BLS	202-501-6795
Carnevale, Sharon	BUS	Employment, State Estimates	BEA	202-254-7703
Carpenter, Douglas	BUS	Reports of Condition and Income	FRS	202-452-2740
Carper, Virginia	BUS	International Trade Policy, U.S.	FRS	202-452-3661
Carsn, Bob	ENV	National Parks, Air Quality	NPS	303-969-2072
Carte, Esther	BUS	Input-Output Tables, Computer Tapes	BEA	202-523-0792
Cartwright, David	BUS	Capital Expenditures	BEA	202-523-0791
Catta, Peter	BUS	Minority Labor Force Data	BLS	202-523-1944
Cevi, Wanda	BUS	County and City Data Books	Census	301-763-1034
Champion, Elinor	BUS	Manufacturing Capacity	Census	301-763-5616
Chan, Valerie	BUS	Foreign Economies, Europe, CIS	FRS	202-452-2375
Chelna, Joseph	BUS	Motor Fuels, Average Retail Prices	BLS	202-272-5080
ChenNash, Elaine	BUS	Data Mgr., Safety and Health Systems	BLS	202-501-6448
Chery, Joseph	BUS	New Investment Survey	BEA	202-523-0654
Christensen, Lee	BUS	Poultry, Prices and Economic Data	ERS	202-219-0714
Cimini, Michael	BUS	Wage Chronologies	BLS	202-523-1320
Clark, Cynthia	BUS	Data Mgr., Agricultural Research	NASS	202-720-4557
Clauson, Annette	BUS	Tobacco Production Costs	ERS	202-219-0890

Part Two: The DATAPHONER

Name	Data Directory	Subject	Agency	Telephone
Cochrane, Nancy	BUS	Agriculture, Eastern Europe	ERS	202-219-0621
Cohany, Sharon	BUS	Veterans Employment	BLS	202-523-1944
Cohen, Barry	BUS	Data Mgr., Economic Programming	Census	301-763-2912
Cohen, Sheryl	BUS	Interest Bearing Notes	FRS	202-452-3471
Cohen, Stephen	BUS	Data Mgr., Wages, Stat. Methods	BLS	202-523-5922
Cole, James	BUS	Data Mgr., Feed Grains Data	ERS	202-219-0840
Cole, Roger	BUS	Capital Adequacy Guidelines	FRS	202-452-2618
Conners, Thomas	BUS	Africa, Asia, Latin Amer. IMF; E-I Bank	FRS	202-452-3639
Cook, William	BUS	Food, Retail Prices	BLS	202-272-5173
Corrado, Carol	BUS	Industrial Output	FRS	202-452-3521
Costello, Brian	BUS	International Machinery Price Indexes	BLS	202-272-5034
Courtland, Sherry	BUS	Data Mgr., Demographic Surveys	Census	301-763-2776
Coveney, Elizabeth	ENV	Ozone Concentration Levels	BNL	516-282-2259
Cowan, Rosemarie	BUS	Population, Special Tabulations	Census	301-763-7947
Coyle, William	BUS	Agriculture, Asia, East	ERS	202-219-0610
Coyne, Joseph	BUS	Discount Rates at Federal Reserve Banks	FRS	202-452-3204
Crabbe, Leland	BUS	Bond Markets, Corporate	FRS	202-452-3022
Crawford, Jeffrey	BUS	Plant and Equipment Expenditures	BEA	202-523-0782
Cromartie, Stella	BUS	Occupational Employment	BLS	202-523-1371
Cron, Robert	ENV	Recreation Sites, U.S. Forests	FS	202-205-1408
Cronkhite, Frederick	BUS	Establishment Survey Benchmark	BLS	202-523-1146
Crutchfield, Steve	ENV	Water Quality	ERS	202-219-0444
Culbreth, Leonard	BUS	Treasury Securities, Interest Rates, Yields	FRS	202-452-3853
Cull, Diana	BUS	Criminal Justice Statistics	Census	301-763-7789
Curran, Thomas	ENV	Ambient Air Quality	OAQPS	919-541-5558
Cypert, Pauline	BUS	Disposable Personal Income	BEA	202-523-0832
Dahl, Thomas	ENV	Wetland Resources	FWS	813-893-3873
Dahmann, Don	BUS	Farm Population	Census	301-763-5158
Daugherty, Arthur	BUS	Agricultural Land Use	ERS	202-219-0424
Davenport, Ronald	BUS	Bank Holding Companies, Domestic	FRS	202-452-3623
Davis, Wanda	BUS	Producer Price Indexes	BLS	202-272-5127
DeAre, Diana	BUS	Migration Data	Census	301-763-3850
DeBraal, Peter	BUS	Farm Real Estate Taxes	ERS	202-219-0425
DeCiccio, Paul	BUS	Foreign Trade Commodity Analysis	Census	301-763-5200
Decker, Patrick	BUS	Foreign Interest and Exchange Rates	FRS	202-452-3314
Decker, Zigmund	BUS	County Business Patterns	Census	301-763-5430
Decorleto, Donna	BUS	T. Bills, Notes, Bonds, Technical Data	FRS	202-452-3954
Delvo, Herman	BUS	Pesticides	ERS	202-219-0456
Dennis, William	BUS	Small Business Employee Benefits	NFIB	202-554-9000
Deven, Richard	BUS	Job Vacancies	BLS	202-523-1694

Index of Data Experts

Name	Data Directory	Subject	Agency	Telephone
Devine, Janice	BUS	Collective Bargaining Settlements	BLS	202-523-1308
DiCesare, Constance	BUS	Data Mgr., BLS Information Services	BLS	202-523-1090
Dietz, Richard	BUS	Consumer Expenditure Data and Tapes	BLS	202-272-5156
Dillard, Fay	BUS	Natural Gas Imports and Exports	FPC	202-586-6181
DiLullo, Anthony	BUS	Balance of Payments, Analysis	BEA	202-523-0621
Dismukes, Robert	BUS	Agricultural Costs and Returns	ERS	202-219-0801
Dixit, Praveen	BUS	Commodity Programs and Policies	ERS	202-219-0632
Dixon, Connie	BUS	Cash Receipts	ERS	202-219-0804
Dixon, Thomas	ENV	Ecological Condition (EMAP)	ORD	202-260-5782
Dobbs, David	BUS	Federal Government Expenditures	BEA	202-523-0744
Donahoe, Gerald	BUS	Saving	BEA	202-219-0669
Dopkowski, Ronald	BUS	Longitudinal Population Surveys	Census	301-763-2767
Douvelis, George	BUS	Farm Output	ERS	202-219-0432
Downs, Bill	BUS	Housing Data	Census	301-763-8553
Doyle, James	BUS	Data Mgr., Computer Services	BEA	202-523-0978
Duewer, Larry	BUS	Price Spreads, Meat	ERS	202-219-0712
Duffield, James	BUS	Farm Labor	ERS	202-219-0033
Dunbar, Ann	BUS	Gross State Product Estimates	BEA	202-523-9180
Dunham, Denis	BUS	Consumer Food Price Index	ERS	202-219-0870
Dunning, Philip	ENV	National Forests	FS	202-205-0843
Dunston, Donna	ENV	Natural Gas Production	EIA	202 586-6135
Durst, Ron	BUS	Farm Taxes	RS	202-219-0896
Dykes, Ellen	BUS	FR Bulletin Articles; Chart Book	FRS	202-452-3952
Edmondson, William	BUS	Economic Linkages to Agriculture	ERS	202-219-0785
Ehemann, Christian	BUS	Exports, Net, U.S.	BEA	202-523-0669
Eldridge, Herb	BUS	Hay, Production and Stocks	NASS	202-720-7621
Elitzak, Howard	BUS	Agricultural Marketing Margins	ERS	202-219-0870
Emerson, Marianne	BUS	Banking Statistics System	FRS	202-452-2045
Emond, Christopher	BUS	International Service Transactions	BEA	202-523-0632
Epps, Walter	BUS	Food Wholesaling	ERS	202-219-0866
Ernst, Lawernce	BUS	Demographic Research	Census	301-763-7880
Evans, Carol	BUS	Wages and Salaries, Subnational	BEA	202-523-0945
Evans, Gardner	ENV	Global Air Quality Monitoring	AREAL	919-541-3887
Evans, Sam	BUS	Futures Markets, Crops	ERS	202-219-0841
Fahim-Nader, Mahnaz	BUS	New Investment Survey	BEA	202-523-0640
Fallert, Richard	BUS	Dairy Biotechnology	ERS	202-219-0710
Fanning, William	BUS	Government Operations	Census	301-763-4403
Farley, Dennis	BUS	Bank Loans	FRS	202-452-3021
Farrell, Michael	BUS	Foreign Trade Data Analysis	Census	301-763-2700
Farrow, Daniel	ENV	Water Pollutants in Coastal Areas	NOS	301-443-0454

Part Two: The DATAPHONER

Name	Data Directory	Subject	Agency	Telephone
Fergus, James	BUS	Housing Starts; Residential Construction	FRS	202-452-2868
Field, Charles	BUS	Wages, State and Local Governments	BLS	202-523-1570
Fischer, Ilene	BUS	International Energy Price Indexes	BLS	202-272-5027
Fitzsimmons, James	BUS	Metro. Areas Concepts and Products	Census	301-763-5158
Flaim, Paul	BUS	Data Mgr., Labor Force Statistics	BLS	202-523-1944
Flood, Thomas	BUS	Current Industrial Reports (CIRs)	Census	301-763-5911
Fogler, Elizabeth	BUS	Balance Sheet, Savings Flows in U.S.	FRS	202-452-3491
Follette, Glen	BUS	Federal Budget National. Debt	FRS	202-452-2448
Fondelier, David	BUS	Housing Completions	Census	301-763-5731
Forstall, Richard	BUS	Metro. Areas Concepts and Products	Census	301-763-5158
Forte, Darlene	BUS	Government Productivity	BLS	202-523-9156
Fouch, Gregory	BUS	Balance of Payments Data	BEA	202-523-0547
Fountain, Melvin	BUS	Occupational Outlook	BLS	202-272-5298
Fox, Douglas	BUS	Data Mgr., Current Business Analysis	BEA	202-523-0697
Frankel, Allen	BUS	Eurocurrency Market, Banking, Taxation	FRS	202-452-3578
Franklin, James	BUS	Input-Output and Industry Data	BLS	202-272-5240
Fraser, Wallace	BUS	Housing Vacancy Data	Census	301-763-8165
Freeman, Richard	BUS	World Payments and Economic Activity	FRS	202-452-2344
Freme, Fred	ENE	Coal, Coke, Employment Production	EIA	202-254-5379
French, Mark	BUS	Energy	FRS	202-452-2348
Frerichs, Stephen	BUS	Organic Farming	ERS	202-219-0401
Friedenberg, Howard	BUS	Regional Economic Situation	BEA	202-523-0979
Fries, Gerhard	BUS	Survey of Consumer Finances	FRS	202-452-2578
Fronzek, Peter	BUS	Housing Markets	Census	301-763-8552
Frumkin, Rob	BUS	International Apparel Price Indexes	BLS	202-272-5028
Fuchs, Doyle	BUS	Hogs, Production and Stocks	NASS	202-720-3106
Fulco, Lawrence	BUS	Productivity, Costs News Releases	BLS	202-523-9261
Fullerton, Howard	BUS	Labor Force Projections	BLS	202-272-5328
Gaddie, Robert	BUS	Food Producers Price Index	BLS	202-272-5210
Gajewski, Greg	BUS	Sustainable Agriculture	ERS	202-219-0883
Galbraith, Karl	BUS	National Defense Purchases	BEA	202-523-3472
Galbreath, Myra	ENV	Municipal Landfills	OSW	202-260-4697
Gallo, Tony	BUS	Food Manufacturing	ERS	202-219-0866
Garrett, Bonnie	BUS	Treasury Securities, Interest Rates, Yields	FRS	202-452-2869
Gates, Jerry	BUS	Population, Privacy Issues	Census	301-763-5062
George, Tommy	ENV	Land Erosion Data	SCS	202-720-6267
Gerduk, Irwin	BUS	Data Mgr., Service Industry Prices	BLS	202-272-5130
Getz, Patricia	BUS	Industrial Classification	BLS	202-523-1172
Gianessi, Leonard	BUS	Herbicide Use	RFF	202-328-5036

Index of Data Experts

Name	Data Directory	Subject	Agency	Telephone
Gibson, J. H.	ENV	Acid Rain, Monitoring Network	GS	303-491-1978
Gibson, Sharon	BUS	Consumer Price Index Data Diskettes	BLS	202-504-2051
Gilbert, Charles	BUS	Manufacturing Capacity	FRS	202-452-3197
Gill, Mohinder	BUS	Energy Statistics	ERS	202-219-0464
Gillum, Gary	BUS	Discounts and Advances, FR Banks	FRS	202-452-3253
Glaser, Lewerne	BUS	Alternative Crops	ERS	202-219-0788
Goldhirsch, Bruce	BUS	Exports, Manufacturing	Census	301-763-1503
Goldstein, Arnold	BUS	Population, Aging	Census	301-763-7883
Gollehon, Noel	BUS	Water and Irrigation	ERS	202-219-0410
Goode, Charles	BUS	Population, World Data	ERS	202-219-0705
Goodman, Nancy	BUS	Annexations	Census	301-763-3827
Gordon, Dale	BUS	Wholesale Trade Inventory	Census	301-763-3916
Govoni, John	BUS	Manufacturing Industry Statistics	Census	301-763-7666
Graham, Claudette	ENE	Imports of Crude Oil	EIA	202-586-9649
Gray, Christine	ENE	Crude Oil Consumption by Pipelines	EIA	202 586-8995
Gray, Fred	BUS	Coffee and Tea Statistics	ERS	202-219-0888
Gray, Kenneth	BUS	Agriculture, Eastern Europe	ERS	202-219-0621
Green, George	BUS	Business Cycle Indicators	BEA	202-523-0800
Green, Gloria	BUS	Employment and Earnings Publication	BLS	202-523-1959
Green, Gordon	BUS	Data Mgr., Government Data	Census	301-763-7366
Greene, Catherine	BUS	Dry Edible Beans	ERS	202-219-0886
Grise, Verner	BUS	Tobacco Commodity Programs	ERS	202-219-0890
Gunderson, Ellis	ENV	Total Diet Study	FDA	202-245-1152
Gustafson, Ron	BUS	Cattle, Prices and Economic Data	ERS	202-219-1286
Hacklander, Duane	BUS	Agricultural Finances	ERS	202-219-0798
Hadlock, Paul	BUS	Occupational Employment Statistics	BLS	202-523-1242
Haidacher, Richard	BUS	Food Demand and Expenditures	ERS	202-219-0870
Hamel, Harvey	BUS	Job Tenure	BLS	202-523-1371
Hamill, Robert	BUS	Congressional District Boundaries	Census	301-763-5720
Hamilton, Howard	BUS	Data Mgr., Business Data	Census	301-763-7564
Hamm, Shannon	BUS	Vegetables, Prices and Economic Data	ERS	202-219-0886
Handy, Charles	BUS	Food Manufacturing and Retailing	ERS	202-219-0866
Hannan, Maureen	BUS	National Information Center	FRS	202-452-3618
Hansen, Kenneth	BUS	Durables, Manufacturing Industry	Census	301-763-7304
Hansen, Kristin	BUS	Place of Birth Statistics	Census	301-763-3850
Harper, Margaret	BUS	NY City Housing and Vacancy Survey	Census	301-763-8552
Harper, Michael	BUS	Capital Measurement	BLS	202-523-6010
Harris-Russell, C.	BUS	Electric Utilities Sales and Purchases	EIA	202-254-5437
Hartman, Frank	BUS	Quarterly Financial Report	Census	301-763-4274
Harvey, Dave	BUS	Aquaculture, Prices and Economic Data	ERS	202-219-0890

Part Two: The DATAPHONER

Name	Data Directory	Subject	Agency	Telephone
Harwood, Joy	BUS	Commodity Programs and Policies	ERS	202-219-0840
Haugen, Steve	BUS	Minimum Wage Data	BLS	202-523-1944
Hawkins, Thomas	BUS	Fish and Wildlife Service Lands	FWS	703-358-1811
Hayghe, Howard	BUS	Family Characteristics of Workers	BLS	202-523-1371
Hazen, Linnea	BUS	County Personal Income	BEA	202-254-6642
Hedges, Brian	BUS	Data Mgr., BLS Statistical Methods	BLS	202-272-2195
Heifner, Richard	BUS	Futures Markets, Agriculture	ERS	202-219-0868
Heimlich, Ralph	BUS	Natural Resource Policy	ERS	202-219-0422
Helkie, William	BUS	International Transactions, U.S.	FRS	202-452-3836
Herman, Shelby	BUS	Price Measures, Chain Price Indexes	BEA	202-523-0828
Herrick, Paul	BUS	Foreign Trade in Machinery and Vehicles	Census	301-763-5200
Herz, Diane	BUS	Working Poor	BLS	202-523-1944
Higbee, Reba	BUS	Foreign Trade Data Inquires	Census	301-763-5140
Higgins, Foster	BUS	Flexible Benefits Programs	HCBS	609-520-2441
Hilaski, Harvey	BUS	Health Studies	BLS	202-272-3459
Hiles, David	BUS	Establishment Survey Data Diskettes	BLS	202-523-1172
Hines, Fred	BUS	Agriculture and Community Linkages	ERS	202-219-0525
Hines, Joseph	BUS	Labor Force Programs	BLS	202-504-2020
Hinton, Dave	BUS	Natural Gas Liquids	EIA	202-586-2990
Hintzman, Kevin	BUS	Fruits and Tree Nuts, Production/Stocks	NASS	202-720-5412
Hirschfeld, Don	BUS	Centers of Population	Census	301-763-5720
Hobbs, Frank	BUS	Africa, Trade Data	Census	301-763-4221
Hofacker, Thomas	ENV	Forest Lands, Insect/Disease Conditions	FS	202-205-1600
Hoff, Fred	BUS	Commodity Programs and Policies	ERS	202-219-0883
Hoff, Gail	BUS	Consumer Expenditure Survey	Census	301-763-2063
Hogan, Howard	BUS	Post-Enumerations Surveys	Census	301-763-1794
Hoheisel, Constance	ENV	Pesticide Information Network (PIN)	OPP	703-305-5455
Holden, Sarah	BUS	Debt, Domestic Non-Financial	FRS	202-452-3483
Holliday, Mark	BUS	Fisheries, Annual World Catch Data	NMFS	301-713-2328
Holmes, Sandra	ENV	Commercial Water Use	GS	703-648-6815
Homer, Laura	BUS	Futures Trading, Securities Credit	FRS	202-452-2781
Honsa, Jeanette	BUS	UN and OECD National Accounts	BEA	202-523-0835
Hooper, Peter	BUS	U.S. International Transactions	FRS	202-452-3426
Hoover, Kent	BUS	Guam Agriculture Census	Census	301-763-8564
Hoppe, Robert	BUS	Income and Poverty, Rural Statistics	ERS	202-219-0547
Horowitz, Karen	BUS	Input-Output Tables, Service Industries	BEA	202-523-3505
Horvath, Francis	BUS	Longitudinal Employment Studies	BLS	202-523-1371
Hoskin, Roger	BUS	Soybeans, Prices and Economic Data	ERS	202-219-0840
Hostetler, John	BUS	Water and Irrigation	ERS	202-219-0410
Houff, James	BUS	Capital Accumulation Benefits	BLS	202-523-8791

Index of Data Experts

Name	Data Directory	Subject	Agency	Telephone
Howell, Craig	BUS	Data Mgr., Industrial Prices Data	BLS	202-272-5113
Howenstine, Barbara	BUS	Computerized Economic Data	BEA	202-523-0777
Howenstine, Ned	BUS	Foreign Direct Investment	BEA	202-523-0650
Hoyle, Kathryn	BLS	Press Officer	BLS	202-523-1913
Hoyle, Linda	BUS	Construction Building Permits	Census	301-763-7244
Hsen, Paul	BUS	Data Mgr., Consumer Expenditures	BLS	202-272-2321
Hubbert, William	ENV	Contaminants in Meat and Poultry	FSIS	202-205-0007
Hurt, Cecilia	BUS	Consumer Leasing	FRS	202-452-2412
Hutchinson, T. Q.	BUS	Transportation, Agriculture	ERS	202-219-0840
Hutton, Linda	BUS	Livestock Statistics	Census	301-763-8569
Ingold, Jane	BUS	School District Data	Census	301-763-3476
Ireland, Oliver	BUS	Res. of Dep. Institutions, Member Banks	FRS	202-452-3625
Isbell, Chuck	BUS	Data on Air Pollution	OAQPS	919-541-5448
Jack, Patricia	BUS	Data Mgr., Price Programs	BLS	202-504-2015
Jackman, Patrick	BUS	Consumer Price Index	BLS	202-272-5160
Jackson, Arnold	BUS	Data Mgr., Decennial Operations	Census	301-763-2682
Jackson, Ethel	BUS	Occupational Injury Data Diskettes	BLS	202-501-6470
Jacobs, Eva	BUS	Data Mgr., Consumer Expenditure Survey	BLS	202-272-5156
Jacobson, Dale	BUS	Residential Construction in MSAs	Census	301-763 7842
Jahnke, Donald	BUS	Crop Statistics	Census	301-763-8567
Jamison, Ellen	BUS	Women in Developing Countries	Census	301-763-4086
Jansen, Anicca	BUS	Local Government Finances, Rural Data	ERS	202-219-0542
Jarema, Frank	BUS	Highway Statistics	FHA	202-366-0160
Jennings, Jack	BUS	Miscellaneous Questions on Banks	FRS	202-452-3053
Jennings, Jerry	BUS	Voting and Registration Data	Census	301-763-4547
Johnson, Doyle	BUS	Floriculture, Prices and Economic Data	ERS	202-219-0884
Johnson, Everette	BUS	Personal Consumption, Autos	BEA	202-523-0807
Johnson, Karen	BUS	Economies, Canada/ Europe/Japan/OECD	FRS	202-452-2345
Johnson, Kenneth	BUS	Regional Economic Projections	BEA	202-523-0971
Johnson, Martha	BUS	Energy Consumption, Commercial Bldgs.	EIA	202-586-1135
Johnson, Melvin	ENE	Electric Generating Plants	EIA	202-254-5665
Johnson, Peter	BUS	International Data Base	Census	301-763-4811
Johnson, Sam	BUS	National Services Information Centers	Census	301-763-1384
Jolley, Charles	BUS	Interest, Subnational Estimates	BEA	202-254-6637
Jones, Selwyn	BUS	Employment	Census	301-763-8574
Jones, Sonja	ENV	Carbon Dioxide, Gas-Related Data	ORNL	615-574-0390
Jorgenson, Harry	BUS	Interest on Deposits/Reserve Requirement	FRS	202-452-3778
Kasper, Marvin	BUS	Data Mgr., International Prices	BLS	202-272-2272
Kasprzyk, Daniel	BUS	Survey, Income Program Participation	Census	301-763-8328
Kaufman, Milton	BUS	Foreign Trade Information	Census	301-763-5940

Part Two: The DATAPHONER

Name	Data Directory	Subject	Agency	Telephone
Kaufman, Phil	BUS	Food Retailing	ERS	202-219-0866
Kazanowski, Edward	BUS	Metals Producer Price Index	BLS	202-272-5204
Kealy, Walter	BUS	Balance of Payment Estimates	BEA	202-523-0625
Keeling, James	ENE	Natural Gas Pipeline Data	FERC	202-586-6107
Keffer, Gerald	BUS	Tax Data	Census	301-763-5356
Kellerman, David	BUS	Federal Government Expenditures	Census	301-763-5276
Key, Greg	BUS	Personal Consumption Expenditures	BEA	202-523-0778
Key, Sidney	BUS	Foreign Banks in U.S.; Internat'l. Banks	FRS	202-452-3522
Kilaru, Vasu	ENV	Toxics Information	OAQPS	919-541-0850
Kim, Po Kyung	BUS	Statistical Services	FRS	202-452-3842
King, Bob	ENV	Ocean Data, Bioaccumulation Data	OW	202-260-7050
King, Bob	ENV	Ocean Data, Benthic Surveys	OW	202-260-7050
King, Glenn	BUS	Statistical Abstract	Census	301-763-5299
Kirsch, Philip	BUS	Data Mgr., Industrial Prices and Relations	BLS	202-272-5182
Klein, Bruce	BUS	Economic Hardship Data	BLS	202-523-1371
Klein, Jerry	ENV	Radioactive Waste and Spent Fuel	ORNL	615-574-6823
Kleweno, Doug	BUS	Farm Prices, Parity Paid	NASS	202-720-4214
Kline, Don	BUS	BHCs, Acquisitions	FRS	202-452-3421
Kobilarcik, Ed	BUS	Decennial Census, 1990 Count	Census	301-763-4894
Kohn, Donald	BUS	Monetary Policy Questions	FRS	202-452-3761
Kohout, Edward	ENV	Emissions, Stationary, Mobile Sources	ANL	708-972-7644
Koopman, Robert	BUS	Agriculture, Eastern Europe	ERS	202-219-0621
Kort, John	BUS	Shift-Share Analysis	BEA	202-523-0946
Kostinsky, Barry	BUS	Soviet Union, General Information	Census	301-763-4022
Kotwas, Gerald	BUS	Foreign Trade Programs	Census	301-763-5333
Kozlow, Ralph	BUS	Services, U.S. Transactions	BEA	202-523-0632
Kriebel, Bertram	BUS	Productivity Data Tapes	BLS	202-523-9261
Kruchten, Tom	BUS	Honey, Production Data	NASS	202-690-4870
Krupa, Ken	BUS	Grassland Pasture and Range	ERS	202-219-0424
Kurtz, Tom	BUS	Farm Labor	NASS	202-690-3228
Kurtzig, Michael	BUS	Agriculture, Africa	ERS	202-219-0680
Lamas, Enrique	BUS	Households, Wealth	Census	301-763-8578
Lambrecht, Ronald	ENE	Vehicle Use	EIA	202-586-4962
Lander, Joel	BUS	Private Pension Funds	FRS	202-452-2227
Landers, Dixon	ENV	Biological Characteristics of Lakes	ERL	503-757-4427
Landes, Rip	BUS	Agriculture, Asia, South	ERS	202-219-0664
Landman, Cheryl	BUS	Census Publications, Decennial	Census	301-763-3938
Lane, Walter	BUS	Data Mgr., Consumer Prices	BLS	202-272-3583
Lange, John	BUS	Cold Storage	NASS	202-382-9185
LaPointe, Tom	ENV	Marine Species	NOS	301-443-0453

Index of Data Experts

Name	Data Directory	Subject	Agency	Telephone
Larson, Odell	BUS	Agriculture, Northern Marianas	Census	301-763-8226
Latham, Roger	BUS	Cotton, Production and Stock Data	NASS	202-720-5944
Latimer, Richard	ENV	Condition of Inland Waters (EMAP)	ORD	401-782-3077
Lavish, Kenneth	BUS	Data Mgr., Durable Goods Prices	BLS	202-272-5115
Lawler, John	BUS	Wool and Mohair, Prices / Economic Data	ERS	202-219-0840
Lawless, Thomas	ENV	Air Quality in Metro Areas	AREAL	919-541-2291
Lawson, Ann	BUS	Balance of Payments	BEA	202-523-0628
Lebow, David	BUS	Labor Markets, Employment and Wages	FRS	202-452-3057
Ledbury, Dan	BUS	Number of Farms	NASS	202-720-1790
Lee, Ronald	BUS	Quarterly Financial Report	Census	301-763-4270
Leitzke, John	ENV	Chemical Hazard Exposure Information	OTS	202-260-3507
Levezey, Janet	BUS	Rice, Prices and Economic Data	ERS	202-219-0840
Levin, Michael	BUS	Population of Outlying Areas	Census	301-763-5134
Levine, Bruce	BUS	Nonfarm Proprietors' Income, Subnational	BEA	202-254-6634
Lichtenstein, Jules	BUS	Health Care Costs in Business	SBA	202-653-6365
Liefer, James	BUS	Farm Economics	Census	301-763-8566
Lienesch, Thomas	BUS	State Econometric Modeling	BEA	202-523-0943
Link, John	BUS	Agriculture, Latin America	ERS	202-219-0660
Lippert, Alice	ENE	Fuel Oil, Distillates	EIA	202-586-9600
Little, Robert	BUS	Aquaculture Production and Stock Data	NASS	202-720-6147
Lord, Ron	BUS	Sugar Commodity Programs, Policies	ERS	202-219-0888
Love, John	ENV	Pesticides	ERS	202-219-0886
Lowe, Jeffrey	BUS	Direct Investment Abroad, U.S. Analysis	BEA	202-523-0649
Lucier, Gary	BUS	Dry Edible Beans	ERS	202-219-0884
Luckett, Charles	BUS	Automobile Loans, Consumer Credit	FRS	202-452-2925
Lumpkin, Stephen	BUS	Mortgage Market Analysis	FRS	202-452-2378
Luster, Geraldine	ENV	Radiation Levels in Air	NAREL	205-270-3433
Lynch, Loretta	BUS	Food Policy, World	ERS	202-219-0689
Mabbs-Zeno, Carl	BUS	Farm Subsidies	ERS	202-219-0631
MacDonald, Brian	BUS	Data Mgr., Occupational / Adm. Data	BLS	202-523-1949
Mack, Juanita	ENE	Natural Gas Pipelines, Financial Data	FERC	202-586-6169
Magiera, Steve	BUS	Finance and Trade Policy	ERS	202-219-0633
Magleby, Richard	BUS	Soil Conservation	ERS	202-219-0435
Maley, Leo	BUS	Input-Output Benchmark Tables	BEA	202-523-0683
Malm, William	ENV	National Parks, Atmospheric Conditions	NPS	303-491-8292
Mangan, Robert	BUS	Nondefense Purchases	BEA	202-523-5017
Mangold, Robert	ENV	Tree Planting, Number of Acres Planted	FS	202-205-1379
Mangold, Robert	BUS	Health Surveys	Census	301-763-5508
Mann, Catherine	BUS	International Trade Policy, U.S.	FRS	202-452-2374
Mannering, Virginia	BUS	Gross Domestic Product	BEA	202-523-0824

Part Two: The DATAPHONER

Name	Data Directory	Subject	Agency	Telephone
Manning, Tom	BUS	Agriculture Census	Census	301-763-1113
Marcinowski, Frank	ENV	Radon Levels	NAREL	202-475-9615
Marcus, Jessie	BUS	Labor Force Diskettes and Tapes	BLS	202-523-1002
Marcus, Sidney	BUS	Finance Statistics	Census	301-763-1386
Mariger, Randall	BUS	Federal Budget Receipts and Outlays	FRS	202-452-3703
Martin, Elizabeth	BUS	Data Mgr., Center for Survey Methods	Census	301-763-3838
Martinson, Michael	BUS	Domestic and Foreign BHCs Operations	FRS	202-452-3640
Masterson, Nancy	ENE	Petroleum Refineries and Blending Plants	EIA	202-586-8393
Masumura, Wilfred	BUS	Industry Statistics	Census	301-763-8574
Mataloni, Raymond	BUS	U.S. Direct Investment Abroad, Analysis	BEA	202-523-3451
Mathews, Ken	BUS	Dairy Product Statistics	ERS	202-219-0770
Mathia, Gene	BUS	Developing Economies	ERS	202-219-0680
Maupin, Ellis	ENE	Underground Gas Storage	EIA	202-586-6178
May, Rick	BUS	Info. System, FR Bank Operations	FRS	202-452-3995
Maynard, Jane	BUS	Housing Inventories	Census	301-763-8551
Mazie, Sara	BUS	Rural Development Information	ERS	202-219-0530
McCarthy, Mary	BUS	Data Mgr., International Price Indexes	BLS	202-272-5026
McClevey, Ken	ENE	Coal Imported by Electric Utilities	EIA	202-254-5655
McCormick, Ian	BUS	Peanuts, World Prices	ERS	202-219-0840
McCormick, William	BUS	Foreign Military Sales	BEA	202-523-0619
McCracken, John	BUS	Data Mgr., International Tech. Assistance	BLS	202-523-9231
McCully, Cathy	BUS	Voting Districts	Census	301-763-3827
McCully, Clint	BUS	Personal Consumption Expendtures	BEA	202-523-0819
McCutcheon, Donna	BUS	Business Owners	Census	301-763-5517
McDaniel, Jacqueline	BUS	National Info., Center; Minority Banks	FRS	202-452-3132
McDivitt, Patrick	BUS	Bills of Exchange	FRS	202-452-3818
McElroy, Bob	BUS	Farm Income	ERS	202-219-0800
McElroy, Michael	BUS	Standard Occupational Classification	BLS	202-523-1684
McGinn, Larry	BUS	Crime Statistics	Census	301-763-1735
McGranaham, David	BUS	Rural Development Information	ERS	202-219-0532
McGuckin, Robert	BUS	Data Mgr., Center for Economic Studies	Census	301-763-2337
McIntire, Robert	BUS	Labor Force Micro- Data Tapes	BLS	202-523-1776
McIntosh, Susan H.	BUS	Debt, Domestic Non-Financial	FRS	202-452-3484
McLaughlin, Mr.	BUS	Financial Reports	FRS	202-452-3829
McLaughlin, Mary	BUS	Bank Rates on Business Loans	FRS	202-452-2259
McMillian, Deborah	BUS	Interest Rates Data; Commercial Paper	FRS	202-452-2851
McNamee, John	BUS	Electric Energy Consumed by Mfgrs.	Census	301-763-5938
McNeil, Jack	BUS	Disability Estimates	Census	301-763-8300
McNulty, Donald	BUS	Data Mgr., Comp. / Working Conditions	BLS	202-523-1228
McPhillips, Regina	BUS	Health Care Cost to Employees	HCFA	301-597-3934

Index of Data Experts

Name	Data Directory	Subject	Agency	Telephone
Mearkle, Haydn	BUS	Foreign Trade Data Services	Census	301-763-7754
Meisenheimer, Joseph	BUS	Educational Attainment	BLS	202-523-1944
Melick, William	BUS	Energy, International	FRS	202-452-2296
Mellor, Earl	BUS	Work Experience	BLS	202-523-1371
Messenbourg, Thomas	BUS	Data Mgr., Economic Census Staff	Census	301-763-7356
Meyer, Allan	BUS	Value of New Construction Put in Place	Census	301-763-5717
Meyer, Leslie	BUS	Cotton, Prices and Economic Data	ERS	202-219-0840
Meyers, Les	BUS	Food Policy	ERS	202-219-0860
Middleton, Millie	BUS	Auto and Consumer Credit Card Data	FRS	202-452-2924
Miller, Catherine	BUS	Data Mgr., Program Development Office	Census	301-763-2758
Miller, Cathy	BUS	Census Tract Boundaries	Census	301-763-3827
Miller, Don	ENE	Electric Utilities, Emissions	ANL	708-972-3946
Miller, Douglas	BUS	Agricultural Data Products	Census	301-763-8561
Miller, Timothy	ENV	Quality of Surface Water	GS	703-648-6868
Milton, Bob	BUS	Farm Prices, Parity and Indexes	NASS	202-720-3570
Minnick, Renee	ENV	National Park Lands	NPS	202-343-3862
Minor, Al	BUS	Health Information	HIAA	202-223-7845
Mintz, Steven	BUS	Electrical Import and Export Data	FE	202-586-9506
Miskura, Susan	BUS	Data Mgr., Year 2000 R and D	Census	301-763-8601
Mitchell, Susanne	BUS	Call Reports; Freedom of Info. Data	FRS	202-452-3684
Mohr, Michael	BUS	Gross Domestic Product by Industry	BEA	202-523-0795
Monaco, Johnny	BUS	Enterprise Statistics	Census	301-763-1758
Monaco, Ralph	BUS	Macroeconomic Conditions	ERS	202-219-0782
Monsour, Nash	BUS	Statistical Research-Economic Programs	Census	301-763-5702
Montague, Barry	ENV	Substance Abuse Data	NIAAA	301-443-3864
Monteiro, Anna	BUS	Minority Banks	FRS	202-452-2948
Montfort, Edward	BUS	American Housing Survey	Census	301-763-8551
Moody, Jack	BUS	Service Industries Census	Census	301-763-7039
Moore, Joel	BUS	Aquaculture Production and Stock Data	NASS	202-720-3244
Morehart, Mitch	BUS	Costs and Returns Data	ERS	202-219-0801
Morgan, Nancy	BUS	Soybeans, World Prices / Economic Data	ERS	202-219-0826
Morisse, Kathryn	BUS	International Trade and BoPs, U.S.	FRS	202-452-3773
Morton, John	BUS	Disability Benefits	BLS	202-523-8791
Moyers, Cora Flaifel	BUS	Electric Power Statistics	FRS	202-452-2476
Murad, Howard	BUS	Trade, Merchandise, International Data	BEA	202-523-0668
Murphy, Robert	ENV	Health of U.S. Population	NCHS	301-436-7068
Musco, Thomas	ENV	Health Information	HIAA	202-223-7863
Musgrave, John	BUS	Capital Consumption Allowance	BEA	202-523-0837
Myers, Stacey	ENE	Imports of Oil, Gas and Petroleum.	EIA	202-586-5130
Nardone, Thomas	BUS	Occupational Data, CPS	BLS	202-523-1944

Part Two: The DATAPHONER

Name	Data Directory	Subject	Agency	Telephone
Natof, Margo	ENE	Origins of Natural Gas Supplies	EIA	202-586-6303
Neef, Arthur	BUS	Data Mgr., Foreign Labor Statistics	BLS	202-523-9291
Nelson, Fred	BUS	Farm Subsidies	ERS	202-219-0689
Nelson, Ken	BUS	Futures Markets, Livestock	ERS	202-219-0712
New, Mark	BUS	Direct Investment Abroad	BEA	202-523-0612
Niemann, Brand	ENV	Environmental Ecological Statistics	CES	202-260-3726
Norfolk, Irving	BUS	Foreign Trade in Chemicals and Sundries	Census	301-763-5186
O'Connell, Martin	BUS	Births and Fertility	Census	301-763-5303
O'Connor, Thomas	ENV	Coastal Areas, Environmental Quality	NOS	301-443-8644
Oberg, Diane	BUS	Foreign Trade, Methodology	Census	301-763-5709
Oliner, Stephen	BUS	Business Fixed Investment	FRS	202-452-3134
Oliveira, Victor	BUS	Farm Labor Market	ERS	202-219-0033
Osborn, Tim	BUS	Natural Resource Policy	ERS	202-219-0405
Otoo, Maria Ward	BUS	Labor Markets, Wages, Unemployment	FRS	202-452-2236
Owens, Marty	BUS	Rice, Production and Stock Data	NASS	202-720-2157
O'Rourke, Michael	BUS	Bank Acquisitions	FRS	202-452-3288
Padgitt, Merritt	BUS	Pesticides	ERS	202-219-0433
Paez, Al	BUS	Decennial Census, General Information	Census	301-763-4251
Paik, Soon	BUS	Leather, Producer Price Index	BLS	202-272-5127
Palumbo, Thomas	BUS	Employment	Census	301-763-8574
Pandolfi, Thomas	ENV	Water Quality, Biological (STORET)	OW	202-260-7030
Parker, Tim	BUS	Employment, Rural Areas	ERS	202-219-0541
Parks, William	BUS	Earnings by Industry	BLS	202-523-1959
Parlett, Ralph	BUS	Consumer Price Index, Food	ERS	202-219-0870
Parrow, Joan	BUS	Price Spreads, Fruits and Vegetables	ERS	202-219-0883
Passero, William	BUS	Consumer Expenditures Data and Tapes	BLS	202-272-5060
Passmore, Wayne	BUS	Certificates of Deposits, Thrifts	FRS	202-452-6432
Paull, Mary	BUS	Coal Consumption, U.S. Mfg. Plants	EIA	202-254-5379
Paulson, Richard	ENV	Hydrologic Hazards and Land Use	GS	703-648-6851
Pautler, P. Charles	BUS	Data Mgr, Agricultural Data	Census	301-763-8555
Payton, M. L.	ENV	Radioactive Waste/Spent Fuel Inventory	OCRWM	202-586-9140
Perez, Agnes	BUS	Poultry, Production and Stock Data	ERS	202-219-0714
Perry, Mike	ENV	Hazardous Waste Sites Data (HAZDAT)	ATSDR	404-639-0720
Peterjohn, Bruce	ENV	Avian Population, Long-Term Trends	FWS	301-498-0330
Peters, Donald	BUS	Purchases, State and Local Government	BEA	202-523-0726
Peterson, Dave	BUS	Farm Land, Acres Irrigated	Census	301-763-8560
Petko, Charles	ENV	Radioactivity Associated with Air	NAREL	205-270-3400
Petrick, Kenneth	BUS	Corporate Profits	BEA	202-523-0888
Petrick, Kenneth	BUS	Corporate Taxes	BEA	202-523-0888
Pettis, Maureen	BUS	Prices in Foreign Countries	BLS	202-523-9301

Index of Data Experts

Name	Data Directory	Subject	Agency	Telephone
Pfuntner, Jordon	BUS	Data Mgr., Occupational Pay and Benefits	BLS	202-523-1246
Pickering, Ranard	ENV	Precipitation, Monitoring Network	GS	703-648-6875
Piencykoski, Ronald	BUS	Inventories, Monthly Retail Trade Data	Census	301-763-5294
Pierce, David	BUS	Seasonal Adjustments	FRS	202-452-3895
Pigler, Carmen	BUS	Regional Input-Output Model (RIMS)	BEA	202-523-0586
Pilli, Anne	ENV	Aquatic Toxicity Data	ERL	218-720-5516
Pilot, Michael	BUS	Occupational Outlook Handbook	BLS	202-272-5382
Pinkos, John	BUS	Industrial Classification, Methodology	BLS	202-523-1636
Planting, Mark	BUS	Input-Output Annual Tables	BEA	202-523-0867
Platto, Kenneth	ENE	Petroleum Product Data	EIA	202-586-6364
Plotkin, Robert	BUS	Futures Trading	FRS	202-452-2782
Podgornik, Guy	BUS	Earnings, State / Local - Data Diskettes	BLS	202-523-1759
Pope, Anne	ENV	Toxic Air Pollutants	OAQPS	919-541-5373
Porter, Gloria	BUS	Decennial Census Tabulations	Census	301-763-4908
Post, Mitchell	BUS	Commercial Paper, Outstanding	FRS	202-452-2720
Powers, Brendan	BUS	Data Mgr., Fed./State Monthly Surveys	BLS	202-523-1001
Pratt, Richard	BUS	Data Mgr., Producer Price Index	BLS	202-272-2196
Preston, John	BUS	Energy Consumption in the Mfg. Sector	EIA	202-586-1128
Preuss, Richard	BUS	Foreign Trade Press Releases	Census	301-763-7754
Price, Charlene	BUS	Food Away from Home	ERS	202-219-0866
Priebe, John	BUS	Industry Statistics	Census	301-763-8574
Promisel, Larry Jay	BUS	Banking, International; OECD	FRS	202-452-3533
Putnam, Judy	BUS	Food Consumption	ERS	202-219-0870
Puzzilla, Kathleen	BUS	Foreign Trade, Methods and Research	Census	301-763-7760
Quarato, Rose	BUS	Zip Code, Geographic Relationships	Census	301-763-4667
Quasney, Adrienne	BUS	Neighborhood Statistics	Census	301-763-4282
Rabil, Floyd	BUS	Retail Food Prices, Monthly Estimates	BLS	202-272-5173
Raddock, Richard	BUS	Capacity Utilization	FRS	202-452-3197
Ramm, Wolfhard	BUS	National Debt, Federal Budget Receipts	FRS	202-452-2381
Rappaport, Barry	BUS	Construction Censuses and Surveys	Census	301-763-5435
Ratliff, Michael	ENV	Forest Utilization, Public Lands Data	BLM	202-208-5717
Rea, John	BUS	Bus.Finance; Capital and Bond Markets	FRS	202-452-3744
Reagan, James	ENV	Acid Precipitation, Status and Trends	AREAL	919-541-4486
Reeder, Richard	BUS	Local Government Finances, Rural Data	ERS	202-219-0542
Reid, Brian	BUS	CDs; Savings Deposits	FRS	202-452-3589
Reilly, John	BUS	Biotechnology	ERS	202-219-0450
Reimund, Donn	BUS	Corporate Farms	ERS	202-219-0522
Reinert, Joseph	ENV	Pesticide Residue on Food	OPP	202-260-7557
Remmers, Janet	ENV	Toxic Organic Compounds in Population	OTS	202-260-1583
Reut, Katrina	BUS	Price Indexes, International	BLS	202-272-5025

Part Two: The DATAPHONER

Name	Data Directory	Subject	Agency	Telephone
Rhoades, Stephen	BUS	Banking Markets; Merges / Acquisitions	FRS	202-452-3906
Ribaudo, Marc	BUS	Natural Resource Policy	ERS	202-219-0444
Ribble, Leigh	BUS	Money Supply, Deposits, Reserves, U.S.	FRS	202-452-2385
Richards, Gregory	BUS	Data Mgr., Periodic BLS Surveys	BLS	202-272-5483
Richards, Ms.	BUS	Analytic Services, FR Bank Performance	FRS	202-452-2705
Richardson, W. Joel	BUS	Data Mgr., Construction Statistics	Census	301-763-7163
Riley, Peter	BUS	Feed Grains, World Prices	ERS	202-219-0824
Riner, Charles	ENE	Fuel Oil Sold by Resellers and Retailers	EIA	202-586-6610
Rives, Sam	ENV	Pesticides and Fertilizer, Amounts Used	NASS	202-720-2324
Roberts, Elizabeth	BUS	Foreign Branches of U.S. Banks	FRS	202-452-3846
Roberts, Tanya	BUS	Food Safety and Quality	ERS	202-219-0864
Robeson, Dwight	BUS	Foreign Trade Procedures	Census	301-763-4340
Robey, Mary	ENV	Navigable Waters, Spills/Potential Spills	CG	202-267-6670
Robinson, Brooks	BUS	Construction Estimates	BEA	202-523-0592
Robinson, Charles	BUS	Business Cycle, Composite Indexes	BEA	202-523-0800
Robinson, Gregg	BUS	Demographic Analysis	Census	301-763-5590
Roff, George	BUS	Residential Improvements and Repairs	Census	301-763-5705
Rogers, Brenda	BUS	Cost for Employee Compensation	BLS	202-523-1165
Rogers, John	BUS	Family Budget	BLS	202-272-5060
Rogich, Donald	ENV	Mine Waste Statistics	BOM	202-634-1187
Rones, Philip	BUS	Occupational Mobility	BLS	202-523-1944
Roningen, Vernon	BUS	Finance and Trade Policy	ERS	202-219-0631
Rosenberg, Elliott	BUS	Price Index Analysis	BLS	202-272-5118
Rosenblum, Larry	BUS	Labor Composition and Hours Worked	BLS	202-523-9261
Rosenthal, Neal	BUS	Data Mgr., Occupational Outlook	BLS	202-272-5382
Rosine, John	BUS	Agriculture	FRS	202-452-2971
Ross, Phil	ENV	Environmental Statistics, National	CES	202-260-2680
Ross, Thomas	ENV	Groundwater Level Data from 200 Sites	GS	703-648-6814
Rossi, Cliff	BUS	Credit and Finance Markets, Rural Data	ERS	202-219-0892
Rowe, John	BUS	Business and Industry Data Centers	Census	301-763-1580
Royce, John	BUS	Transportation Price Indexes	BLS	202-272-5131
Rubin, Laura	BUS	State and Local Sector Fiscal Data	FRS	202-452-3130
Runyan, Ruth	BUS	Manufacturing Inventories	Census	301-763-2502
Runyon, Jack	BUS	Farm Labor Laws	ERS	202-219-0932
Russell, Anne	BUS	Zip Code Economic Data	Census	301-763-7038
Russell, Charlene	ENE	Electric Utilities, Financial Data	EIA	202-254-5437
Rutledge, Gary	BUS	Environmental Studies	BEA	202-523-0687
Ryan, Jim	BUS	Credit and Financial Markets	ERS	202-219-0798
Ryback, William	BUS	International Banking Operations	FRS	202-452-2722
Salopek, Phil	BUS	Journey to Work Data	Census	301-763-3850

Index of Data Experts

Name	Data Directory	Subject	Agency	Telephone
Saluter, Arlene	BUS	Living Arrangements	Census	301-763-7987
Sanford, Scott	BUS	Peanuts, World Prices and Economic Data	ERS	202-219-0840
Sasnett, Samuel	ENV	Toxic Releases, Atmospheric Emissions	OTS	202-260-1821
Saunders, Norman	BUS	Economic Projections	BLS	202-272-5248
Savage, Donald	BUS	Interstate Banking	FRS	202-452-2613
Schmidt, Catherine	BUS	Benefits for Part-Time Workers	HA	708-295-5000
Schmitt, Christopher	ENV	Contaminants, Fish and Wildlife	FWS	314-875-1800
Schneider, Paula	BUS	Data Mgr., Population Data	Census	301-763-7646
Schoenfeld, Michael	BUS	Municipal Securities Dealer Banks	FRS	202-452-2781
Scholl, Russell	BUS	Capital Transactions, Private	BEA	202-523-0603
Schuchardt, Rick	BUS	Honey, Price Data	NASS	202-690-3236
Scott, Elizabeth	ENE	Cost of Crude Oil Purchased by Refiners	EIA	202-586-1258
Sears, David	BUS	Community Development	ERS	202-219-0544
Searson, Michael	BUS	Business Establishment List	BLS	202-523-6462
Secquety, Roger	ENE	Revenues / Expenses of Electric Utilities	FERC	202-254-5556
Seifert, Mary Lee	BUS	Earnings, Establishment Survey	BLS	202-523-1172
Sensenig, Arthur	BUS	Employee Compensation	BEA	202-523-0809
Seskin, Eugene	BUS	National Income	BEA	202-523-0848
Sexton, Cecil	ENV	Contaminant Levels, Drinking Water	OGDW	202-260-7276
Shafer, Ronald	ENV	Environmental Economic Statistics	CES	202-260-6966
Shagam, Shayle	BUS	Livestock, World Prices	ERS	202-219-0767
Shapiro, Janet	BUS	Manufacturing Pollution Abatement	Census	301-763-1755
Shapouri, H.	BUS	Livestock Production Costs	ERS	202-219-0770
Share, David	BUS	Data Mgr., International Price Systems	BLS	202-504-2158
Sharples, Jerry	BUS	Finance and Trade Policy	ERS	202-219-0791
Shelly, Wayne	BUS	Employment Cost Index	BLS	202-523-1165
Shields, Dennis	BUS	Fruits and Tree Nuts, Prices	ERS	202-219-0884
Shipler, Glenda	BUS	Cattle, Production and Stocks	NASS	202-720-3040
Shipp, Kenneth	BUS	Employment Data by State and Area	BLS	202-523-1227
Shoemaker, Dennis	BUS	Communications Statistics	Census	301-763-2662
Short, Olivia	ENV	Fish and Wildlife Service Lands	FWS	703-358-1811
Short, Sara	BUS	Dairy Product Prices and Economic Data	ERS	202-219-0769
Siegal, Lewis	BUS	Mass Layoff Statistics	BLS	202-523-1807
Siegenthaler, Vaughn	BUS	Wheat, Production and Stocks	NASS	202-720-8068
Siegman, Charles	BUS	Bank for International Settlements	FRS	202-452-3308
Simon, Timothy	BUS	Open Market Operations; Govt. Finance	FRS	202-452-2383
Simone, Mark	BUS	Agriculture, Canada	ERS	202-219-0610
Simpson, Linda	BUS	Sheep, Production and Stocks	NASS	202-720-3578
Sinclair, James	BUS	Electric Machinery Producer Price Index	BLS	202-272-5052
Singh, Raj	BUS	SIPP Statistical Methods	Census	301-763-7944

Name	Data Directory	Subject	Agency	Telephone
Skinner, Robert	BUS	Cotton, Prices and Other Economic Data	ERS	202-219-0841
Slaughter, Eric	ENV	Bacterial Water Quality	NOS	301-443-8843
Small, Richard	BUS	Employees Rights to Benefits	PWBA	202-523-8776
Smallwood, Dave	BUS	Food Assistance and Nutrition	ERS	202-219-0864
Smallwood, Dave	BUS	Food Policy	ERS	202-219-0864
Smith, Denise	BUS	Group Quarters Population	Census	301-763-7883
Smith, George	BUS	Farm Income	BEA	202-523-0821
Smith, Luther	ENE	Nuclear Reactor Construction Data	EIA	202-254-5565
Smith, Mark	BUS	Export Commodity Programs, Policies	ERS	202-219-0821
Smith, Ralph	BUS	Intl. Financial Operations; Swap Network	FRS	202-452-3712
Smith, Richard	ENV	Groundwater Characteristics	GS	703-648-6870
Smith, Susan	ENE	Public/Investor-Owned Electric Utilities	EIA	202-254-5668
Smith, William	BUS	Administrative Salaries	BLS	202-523-1570
Smoler, Anne	BUS	Housing Markets, Finances, Absorption	Census	301-763-8552
Sorrentino, Constance	BUS	Labor Force in Foreign Countries	BLS	202-523-9301
Speaker, Robert	BUS	Apportionment	Census	301-763-7962
Speight, Genevieve	BUS	Survey Operations	Census	301-763-7783
Spinelli, Felix	BUS	Hogs, Prices and Economic Data	ERS	202-219-0713
Stallings, Dave	BUS	World Food Demand and Expenditures	ERS	202-219-0708
Stam, Jerry	BUS	Credit and Financial Markets	ERS	202-219-0892
Steffeck, Donald	ENV	Contaminants, Fish and Birds	FWS	703-358-2148
Stekler, Lois	BUS	Eurobond Mkt, Intl. Capital Transactions	FRS	202-452-3716
Stelluto, George	BUS	Data Mgr., Working Conditions	BLS	202-272-1382
Stevens, Alan	BUS	Government Employment	Census	301-763-5086
Stewart, Louisa	BUS	State Boundary Certification	Census	301-763-3827
Still, Gloria	BUS	Foreign Trade in Food, Animals, Wood	Census	301-763-5211
Stiller, Jean	BUS	Inventory / Sales Ratios	BEA	202-523-6585
Stillman, Richard	BUS	Sheep, Prices and Economic Data	ERS	202-219-0714
Stinson, John	BUS	Earnings by Industry	BLS	202-523-1959
Stoddard, John	ENV	Acid Rain, Monitoring Project (ANC)	CERL	503-757-4427
Storch, Gerald	BUS	Industrial Production Index	FRS	202-452-2932
Strickland, Roger	BUS	Farm Income	ERS	202-219-0804
Struble, Frederick	BUS	Resolution Trust Corp. Oversight Board	FRS	202-452-3794
Struckmeyer, Sandy	BUS	National Income, Price and Inflation	FRS	202-452-3090
Suarez, Nydia	BUS	Food Aid Programs	ERS	202-219-0821
Sullivan, David	BUS	Expenditures, State/Local Government	BEA	202-523-0725
Sullivan, Pat	BUS	Credit and Financial Markets	ERS	202-219-0719
Summers, Jacob	ENV	Pollutants, Air	OAQPS	919-541-5695
Sussan, Sidney	BUS	Thrift Acquisitions, BHC Mergers	FRS	202-452-2638
Sutton, Cynthia	BUS	International Information Center	FRS	202-452-3411

Index of Data Experts

Name	Data Directory	Subject	Agency	Telephone
Swaim, Paul	BUS	Employment, Rural Areas	ERS	202-219-0552
Swanson, Linda	BUS	Population	ERS	202-219-0535
Swieczkowski, Gloria	BUS	Commuting, Journey to Work	Census	301-763-3850
Taeuber, Cynthia	BUS	Homeless Population	Census	301-763-7883
Tague, Norman	BUS	Foreign Trade, Transportation	Census	301-763-7770
Talbert, Cathy	BUS	Redistricting	Census	301-763-4070
Taylor, Harold	BUS	Fertilizer	ERS	202-219-0164
Teigan, Lloyd,	BUS	Weather	ERS	202-219-0705
Teplin, Albert Mack	BUS	Financial Assets and Liabilities in U.S.	FRS	202-452-3482
Terwilliger, Yvonne	BUS	Unemployment Insurance Claimants	BLS	202-523-1002
Thomas, Gregory	BUS	Transactions, Government	BEA	202-523-0615
Thompson, John	BUS	Data Mgr., Statistical Support	Census	301-763-4072
Thompson, Wendel	ENE	Energy Use in the Residential Sector	EIA	202-586-1119
Tibbetts, Thomas	BUS	Data Mgr., Industrial Prices/Price Indexes	BLS	202-272-5110
Tice, Thomas	BUS	Feed Grains, Prices and Economic Data	ERS	202-219-0840
Tichler, Joyce	ENV	Nuclear Power Plants, Releases	BNL	516-282-3801
Tinsley, P. A.	BUS	Monetary Policy Stabilization	FRS	202-452-2438
Tormey, George	BUS	Foreign Trade	Census	301-763-7750
Torrey, Barbara Boyle	BUS	Data Mgr., International Research Center	Census	301-763-2870
Tortora, Robert	BUS	Data Mgr., Statistical Research	Census	301-763-3807
Toscano, Guy	BUS	Occupational Injuries, Fatal	BLS	202-501-6459
Towers, Sharon	BUS	Agriculture, General Information	Census	800-523-3215
Trimble, John	BUS	Wholesale Trade Census	Census	301-763-5281
Trott, Edward	BUS	BEA Economic Areas	BEA	202-523-0973
True, Irving	BUS	Retail Trade, Monthly Report	Census	301-763-7128
Tryon, Ralph	BUS	Multi-Country Trade and Finance Model	FRS	202-452-2368
Tschetter, Marybeth	BUS	Data Mgr., Price Index Methods	BLS	202-272-5170
Tsehaye, Benyam	BUS	Contributions, Federal Government	BEA	202-523-0885
Tuan, Francis	BUS	Agriculture, China	ERS	202-219-0626
Tucker, Robert	BUS	Current Population Survey	Census	301-763-2773
Tucker, Ronald	BUS	Population Estimates	Census	301-763-2773
Tupek, Alan	BUS	Data Mgr., Employment / Unemployment	BLS	202-523-1695
Turner, Christopher	BUS	Bond Markets, Municipal, Tax-exempt	FRS	202-452-2983
Turner, Delores	BUS	Industry-Occupation Employment Matrix	BLS	202-272-5383
Turner, Marshall	BUS	Data Mgr., Data User Services	Census	301-763-5820
Uglow, David	BUS	Data Mgr., Consumer Price Index	BLS	202-272-2323
Ulmer, Mark	BUS	Real Earnings Data	BLS	202-523-1172
Urban, Francis	BUS	World Farm Output	ERS	202-219-0717
Van Giezen, Robert	BUS	Service Contract Act Wage Surveys	BLS	202-523-1536
Van Horne, Merle	ENV	Wild and Scenic Rivers Data	NPS	202-343-3765

Part Two: The DATAPHONER

Name	Data Directory	Subject	Agency	Telephone
Van Lahr, Charles	BUS	Feed Grains, Production and Stocks	NASS	202-720-7369
Van Remortel, Rick	ENV	Ecological Condition of Forests (EMAP)	ORD	702-734-3295
Vandeman, Ann	ENV	Pesticides	ERS	202-219-0433
Vanderberry, Herb	BUS	Soybeans and Sunflowers, Production	NASS	202-720-9526
Venning, Alvin	BUS	Industry Classification Information	Census	301-763-1935
Vesterby, Marlow	BUS	Farm Machinery	ERS	202-219-0422
Vincent, Susan	BUS	Research Library (FRS)	FRS	202-452-3398
Vinnedge, Donald	BUS	Member Banks, Trust Activities	FRS	202-452-2717
Visnansky, Bill	BUS	Construction Statistics	Census	301-763-7546
Vogel, Frederic	BUS	Data Mgr., Agricultural Estimates	NASS	202-720-3896
Wagner, Janice	BUS	Acid Deposition, Causes and Effects	AEERL	919-541-1818
Waite, Preston	BUS	Data Mgr., Census Statistical Methods	Census	301-763-2672
Walker, Patricia	BUS	Direct Investment Abroad	BEA	202-523-0612
Wallace, Charles	BUS	Labor Related Information	BLS	202-523-1208
Walraven, Nicholas	BUS	Agricultural Credit	FRS	202-452-2655
Walter, Bruce	BUS	Foreign Trade, Methodology	Census	301-763-7020
Wapshur, Ernie	BUS	Congressional District Locations	Census	301-763-5692
Warden, Thomas	BUS	Exports	ERS	202-219-0822
Wascher, William	BUS	Labor Markets, Wages, Unemployment	FRS	202-452-2812
Wassom, Molly	BUS	Thrift Acquisitions	FRS	202-452-2305
Watts, Patricia	BUS	Transportation, International	BEA	202-523-0611
Way, Peter	BUS	International Health Statistics	Census	301-763-4086
Weadock, Teresa	BUS	Interest Income	BEA	202-523-0833
Webb, Michael	BUS	Budget, Cyclically-Adjusted	BEA	202-523-3470
Weber, William	BUS	Data Mgr., Safety Health, Programs	BLS	202-501-6468
Weeden, George	BUS	Price Indexes, Service Industry	BLS	202-272-5130
Weinberg, Daniel	BUS	Data Mgr., Housing and Household Data	Census	301-763-8550
Werking Jr., George	BUS	Data Mgr., Employment Statistics	BLS	202-523-1446
Westcott, Paul	BUS	Crop Commodity Programs and Policies	ERS	202-219-0840
Whetzel, Frederick	BUS	Mortgage Market Data, Housing Starts	FRS	202-452-3094
Whichard, Obie	BUS	Services, U.S. Transactions, Analysis	BEA	202-523-0646
Whiston, Isabelle	BUS	State Personal Income, Quarterly Data	BEA	202-254-6672
White, Don	BUS	Health Information	HIAA	202-223-7782
White-Dubose, Gwen	BUS	Federal Reserve Bulletin Tables	FRS	202-452-3567
Whitener, Leslie	BUS	Farm Labor	ERS	202-219-0932
Whitener, Leslie	BUS	Labor Market, Farm	ERS	202-219-0932
Whitmore, Robert	BUS	OSHA Recordkeeping Requirements	BLS	202-272-3462
Whitton, Carolyn	BUS	Cotton, World Prices and Economic Data	ERS	202-219-0824
Wiatrowski, William	BUS	Disability Benefits and Paid Leave	BLS	202-523-8791

Index of Data Experts

Name	Data Directory	Subject	Agency	Telephone
Wilen, Bill	ENV	Wetland Resource	FWS	703-358-2201
Williams, Harry	BUS	Industry Wage Surveys	BLS	202-523-1667
Williams, Janet	BUS	Data Mgr., Consumer Price Index	BLS	202-272-2281
Williams, Robert	BUS	Cash Receipts	ERS	202-219-0804
Williamson, Robert	ENV	Forests, National, Data on Grazing	FS	202-205-1460
Wise, Stephen	ENV	Marine Mammals (Alaskan)	NIST	301-975-3112
Witucki, Larry	BUS	Poultry, Prices and Economic Data	ERS	202-219-0766
Wolffrum, Margaret	BUS	OTC Margin Stock List	FRS	202-452-2781
Wolotira, Robert	ENV	Seabird Colonies	NOS	301-443-0453
Woltman, Henry	BUS	Sampling Methods, Decennial Census	Census	301-763-5987
Won, Gregory	BUS	Computer Price Index	BEA	202-523-5421
Wood, John	ENE	Proved Reserves/Production of Crude Oil	EIA	214-767-2200
Wood Jr., Donald	BUS	Data Mgr., Employment Cost Trends	BLS	202-523-1160
Woodrow, Karen	BUS	Emigration	Census	301-763-5590
Woods, Charles	BUS	Data Mgr., Foreign Trade Data Systems	Census	301-763-7982
Worden, Gaylord	BUS	Data Mgr., Industry Data	Census	301-763-5850
Wright, Stephen	BUS	Data Mgr., Consumer Prices/Consumption	BLS	202-272-5002
Wulf, Henry	BUS	Financial Data, Government	Census	301-763-7664
Wunderlich, Gene	BUS	Farm Land Ownership and Tenure	ERS	202-219-0425
Wysocki, Adam	BUS	Foreign Trade in Textiles	Census	301-763-5138
Young, Edwin	BUS	Rice, Prices and Economic Data	ERS	202-219-0840
Young, Mary	BUS	Business Cycle, Statistical Series	BEA	202-523-0500
Zabelsky, Tom	BUS	Motor Freight Transportation	Census	301-763-1725
Zabronsky, Daniel	BUS	Regional Income Residence Adjustments	BEA	202-254-6639
Zampogna, Michael	BUS	Industry Data, Nondurables Mfg.	Census	301-763-2510
Zarrett, Paul	BUS	Financial Data, Quarterly Estimates	Census	301-763-2718
Zavrel, James	BUS	Proprietors' Income, Farm	BEA	202-254-6638
Zeimetz, Kathryn	BUS	Commonwealth of Independent States	ERS	202-219-0624
Zeisset, Paul	BUS	Economic Census Products	Census	301-763-1792
Zellers, Phillip	BUS	Data Mgr., Agricultural Data Systems	NASS	202-720-2984
Zepp, Glenn	BUS	Potatoes, Prices and Economic Data	ERS	202-219-0883
Ziegler, Martin	BUS	Data Mgr., Employment / Unemployment	BLS	202-523-1919
Ziemer, Richard	BUS	Data Mgr., Government Data	BEA	202-523-0715
Zieschang, Kimberly	BUS	Data Mgr., Price/Index Number Research	BLS	202-272-5096

Subject Index

Subject Index

A
Administrative Record Data 21, 39
Air Quality 191
American Statistics Index (ASI) 29
Annual Rates 39

B
Benchmark Data 39
Benefit-Cost Analysis 39
Bureau of Economic Analysis (BEA) 7
Bureau of Economic Analysis State User Groups 128
Bureau of Labor Statistics (BLS) 8
Bureau of Labor Statistics Regional Offices 154
Bureau of the Census (Census) 7
Business Cycle (Coincident) Indicators 39
Business Cycle (Composite Indexes) Indicators 39
Business Cycle (Lagging) Indicators 40
Business Cycle (Leading) Indicators 40
Business Sector Productivity Measure 40

C
Census Bureau's National Clearing House for Census Data Services 165
Census Bureau Business and Industry Data Centers (BIDCs) 124
Census Bureau State Data Centers 118
Census Catalog and Guide 9
Coal and Coke 209
Commercial Data Services 165
Condition of Education 31
Constant Dollar Estimates 40
Construction Industries 40
Consumer Price Index (CPI) 40
Contacts for Major Data Factory Publications 11

251

Corporate Profits 41
Crude Oil 209
Current Population Survey (CPS) 41

D

Data Factory Publications 11
Data Series 41
Demand 41
Diffusion Index 41
Digest of Education Statistics 31
Disposable Personal Income 41

E

Econometric Model 42
Economic Research Service (ERS) 8
Economic Time Series 42
Educational Data Publications 31
Effective Demand 42
Electric Plants and Electric Utilities 210
Employment Cost Index (ECI) 42
Energy Data Directory 203
Energy Data Sources 207
Energy Data, Uses 213
Energy Information Administration (EIA) 9
Energy Sources, Other 211
Environmental Data Directory 183
Environmental Data Sources, National and International 186
Environmental Data, Other 201
Establishment Survey Data 21
Export Promotion Data System (NTDB) 13
Exports 42
Exports (Domestic) 42
Exports (Foreign or Re-Exports) 43
Extrapolation 23, 43

F

Federal Depository Libraries 5
Federal Reserve Bulletin 28
Federal Reserve System 8
Finance Industries 43
Fixed-Weighted Price Index 43
Foreign Trade Zones 43

G

Government Printing Office Bookstores	34
Gross	43
Gross Domestic Product (GDP)	44
Gross National Product (GNP)	44
Guide to Selected Environmental Statistics in the U.S. Government	186

H

Hard data	23
Housing and Demographic Analysis Division	9

I

Implicit Price Deflator	44
Imports	44
Imputations	45
Income	45
Index Numbers	45
Index of Federal Government Data Experts	223
Index of International Statistics (IIS)	29
Industry Productivity	45
INFOTERRA Network	186
INFOTERRA/USA	186
Input-Output Tables	46
Insurance Business	46
International Economic Data System (NTDB)	13
International Statistics	29
Interpolation	23, 46
Inventory Valuation Adjustment and Profits	46

L

Land Use	193

M

Manufacturing Industries	46
Mean	46
Median	47
Merchandise Trade Balance	47
Merchandise Trade Data	47
Methodology	20
Mineral Industries	47
Monthly Labor Review	27
Multi-Factor Productivity	47

N

National Agricultural Statistical Service (NASS)	8
National Agricultural Statistics Service State Statisticians' Offices	159
National Energy Information Center (NEIC)	205
National Income	48
National Income and Product Accounts	48
National Statistics	29
National Technical Information Service (NTIS)	11
National Trade Data Bank (NTDB)	13
Natural Gas	210
Net	48
Net Exports of Goods and Services	48
Net National Product	48
Nominal Dollar Estimates	48
Non-Governmental Statistics	30
Nonsampling Error	49
NTDB's Export Promotion Data System	13
NTDB's International Economic Data System	13

O

Office of Business Analysis	9, 13
Oil and Petroleum Products	211

P

Personal Consumption Expenditures	49
Personal Income	49
Personal Outlays	49
Personal Savings	49
Private Households	49
Producer Price Index	50
Productivity Measures	50
Public Administration	50
Publications, Economic and Financial	27

Q

Quarterly Financial Report of Manufacturing, Mining, and Trade Corporations	50

R

Real Estate Business	50
Regression Analysis	51
Regulatory Report Data	21, 51
Retail Trade Establishments	51
Revisions	19

S

Samples	51
Sampling Error	51
Seasonal Adjustments	52
Seasonally Adjusted Annual Rates	52
Service Industries	52
Service Sector	52
Services	52
SIC, Standard Industrial Classification	52
SIPP, Survey of Income and Program Participation	53
SITC, Standard International Trade Classification	53
Soft data	23
Solid and Toxic Waste	196
Sources of Data	20
State and Regional Data Centers	115
Statistical (National) Discrepancy	53
Statistical Abstract of the United States	33
Statistical Reference Index (SRI)	30
Statistics of Income Division	9
Supply	53
Survey of Current Business	27

T

Technological Unemployment	53
Terms, Concepts and Measures	37
The Condition of Education	31
Trade Associations	16
Transfer Payments	54
Transportation, Communications, Electric, Gas, and Sanitary Services	54

U

Unemployment	54
Unemployment (Total)	54

W

Water Quality and Water Use	197
Wholesale Trade Businesses	54

About the Authors

Edwin J. Coleman is president of Coleman Consultants, a Maryland-based research and economic data consulting firm. From 1955 to 1985, he managed the U.S. Department of Commerce, Bureau of Economic Analysis's regional economic measurement programs. He was responsible for preparing state and local area estimates of personal income published in the Survey of Current Business and had the principal responsibility for the preparation and publication of State Personal Income: Estimates for 1929-1982—A Statement of Sources and Methods. Coleman has worked closely with more than forty federal statistical agencies, monitoring the availability of economic data sources used to provide the central statistical picture of the nation's economy. He also serves as a consultant to business, government and universities on the use of state and local income measures and regional foreign direct investment.

Ronald A. Morse has focused on trade and competitiveness issues in Asia and America for twenty years. He is currently president of Annapolis International, a Maryland fund raising and research consulting firm specializing in relations with Japan and Northeast Asia. From 1974 to 1981 he served in the U.S. Departments of Defense, State and Energy. He was the fund raiser for the Woodrow Wilson International Center for Scholars and directed the Wilson Center's Asia Program (1981-1988). Morse joined the Library of Congress in 1988 to establish the Development Office and serve as a special assistant for private sector programs to the Librarian of Congress. In 1992, he published a fund raising book, Inside Japanese Support (Taft Group). Morse next became executive vice president at the Economic Strategy Institute, where he helped co-author the book Powernomics: Economics and Strategy after the Cold War (Madison Books). He established Annapolis International in 1991. He serves on a number of advisory boards and is a regular contributor to several publications. Morse has a doctorate in Japanese studies from Princeton University.

ORDER FORM
Detach and send with payment

DATA: WHERE IT IS AND HOW TO GET IT

Quantity ordered _____ X $24.95 $ _____

Maryland Residents: Add 5% Sales Tax $ _____

Shipping and Handling: First Book: $4.00 $ _____

Each Additional Book: Add $1.50 $ _____

Foreign Orders: Add $25.00 $ _____

Note—Books are available at quantity discounts for bulk purchases.   **TOTAL** $ _____

PAYMENT INFORMATION:

☐ Enclosed is a check payable to Coleman / Morse Associates Ltd.

☐ Please charge the amount of $ _____ to my: (check one)

☐ VISA   ☐ MasterCard   ☐ AMERICAN EXPRESS

Card # |__|__|__|__|__|__|__|__|__|__|__|__|__|__|__|__|__|__|__| Exp. Date ___ / ___

Signature _____

SHIP MY COPY OF DATA: WHERE IT IS AND HOW TO GET IT TO:

Name _____

Address _____

City _____ State _____ Zip _____

Daytime Telephone _____

SEND ORDER PAYABLE TO:

Coleman / Morse Associates Ltd.
Suite 231
1290 Bay Dale Drive
Arnold, MD 21012-2325